Current Clinical Urology

For further volumes:
http://www.springer.com/series/7635

Wayne J.G. Hellstrom
Editor

Androgen Deficiency and Testosterone Replacement

Current Controversies and Strategies

Humana Press

Editor
Wayne J.G. Hellstrom, MD, FACS
Department of Urology
Tulane University Medical Center
New Orleans, LA, USA

ISBN 978-1-62703-178-3 ISBN 978-1-62703-179-0 (eBook)
DOI 10.1007/978-1-62703-179-0
Springer New York Heidelberg Dordrecht London

Library of Congress Control Number: 2012951129

Printed on acid-free paper

Humana Press is a brand of Springer
Springer is part of Springer Science+Business Media (www.springer.com)

Preface

The purpose of this book is to provide professionals who study and treat men with androgen deficiency, the latest developments, evolving concepts, and strategies for the treatment of this ever-growing field.

Over the last decade, there has been double-digit growth in the industry that treats male hypogonadism with androgen replacement. Still, the majority of men who suffer with this condition are left untreated. More importantly, there is a paucity of long-term randomized controlled clinical trials in middle-aged and elderly men from which to base our treatment decisions. The Institute of Medicine (IOM) initiated such a plea in 2004 and only recently has the National Institute of Health (NIH) instituted a multicentered study.

Often, testosterone replacement is given a bad rap as a "sex" drug or as an unscientific attempt to counter the aging process. Actually, testosterone is a ubiquitous hormone that positively influences cells, tissues, and organs throughout the body. Each chapter in this book is written by experts in their specific field and addresses different aspects of androgen deficiency and replacement therapy. Short- and long-term benefits, risks, and safety issues are all addressed. While much has been exposed from a current knowledge basis, one recognizes the enormous amount that still needs to be studied scientifically and translated into clinical practice.

I wish to thank all of my colleagues and contributors who pleasantly and promptly rose to the challenge and provided their expert chapters, as well as the professional and positive attitude by Scott Thompson, my editorial assistant, Landon Trost and Michael D. Sova from Springer Science who provided support and tedious dedication to this successful endeavor.

New Orleans, LA, USA Wayne J.G. Hellstrom, MD, FACS

Contents

Contributors

Frances Alba, MD Department of Urology, University of Texas Health Science Center, Houston, TX, USA

Jane Ashby, MBBS, MRCP Department of HIV and Sexual Health, Berkshire Healthcare NHS Foundation Trust, Slough, UK

George C. Bailey Department of Urology, Mayo Clinic, Rochester, MN, USA

Gina Cambareri, MD Division of Urology, Department of Surgery, UMDNJ New Jersey Medical School, Newark, NJ, USA

Giovanni Corona, MD, PhD Andrology and Sexual Medicine Unit, Department of Clinical Physiopathology, University of Florence, Florence, Italy

Endocrinology Unit, Maggiore-Bellaria Hospital, Bologna, Italy

Craig F. Donatucci, MD Department of Urology, Duke University, Durham, NC, USA

Ahmet Gokce, MD Department of Urology, Tulane University School of Medicine, New Orleans, LA, USA

David Goldmeier, MD, FRCP Jane Wadsworth Clinic, Imperial College NHS, London, UK

André T. Guay, MD, FACP, FACE Center for Sexual Function, Harvard Medical School, Peabody, MA, USA

Wayne J.G. Hellstrom, MD, FACS Department of Urology, Tulane University Medical Center, New Orleans, LA, USA

Mohit Khera, MD, MBA, MPH Scott Department of Urology, Baylor College of Medicine, Houston, TX, USA

Andrew C. Kramer, MD Department of Surgery, The University of Maryland School of Medicine, Baltimore, MA, USA

Tracey L. Krupski, MD, MPH Department of Urology, University of Virginia, Charlottesville, VA, USA

Gregory J. Lowe, MD Department of Urology, Ohio State University Medical Center, Columbus, OH, USA

Mario Maggi, MD Andrology and Sexual Medicine Unit, Department of Clinical Physiopathology, University of Florence, Florence, Italy

Erin R. McNamara, BA, MS, MD Division of Urology, Department of Surgery, Duke University Medical Center, Durham, NC, USA

Akanksha Mehta, MD Department of Urology, Weill-Cornell Medical College, New York, NY, USA

Gregory C. Mitchell, MS Department of Urology, Tulane University Medical Center, New Orleans, LA, USA

Darius A. Paduch, MD, PhD Consulting Research Services, Inc., Red Bank, NJ, USA

Department of Urology and Reproductive Medicine, Weill-Cornell Medical College, New York, NY, USA

Giulia Rastrelli, MD Andrology and Sexual Medicine Unit, Department of Clinical Physiopathology, University of Florence, Florence, Italy

Claudio A. Romero, MD Department of Urology, University of Texas Health Science Center, Houston, TX, USA

Sherry S. Ross, MD Department of Surgery, Duke University Medical Center, Durham, NC, USA

Hossein Sadeghi-Nejad, MD, FACS Division of Urology, Department of Surgery, UMDNJ New Jersey Medical School, Newark, NJ, USA

Hackensack University Medical Center, Hackensack, NJ, USA

Ege Can Serefoglu, MD Department of Urology, Tulane University Medical Center, New Orleans, LA, USA

Run Wang, MD, FACS Department of Urology, University of Texas Medical School at Houston and MD Anderson Cancer Center, Houston, TX, USA

Chapter 1
The Institute of Medicine White Paper on Testosterone: Current Perspective

Craig F. Donatucci

Good health and normal sexual function in males require functioning testicles, a fact known since antiquity. It is only more recently that the source of that effect was attributed to testosterone [1]. Several centuries ago, Arnould Berthold castrated a rooster as part of an experiment, which resulted in loss of the physical signs of masculinization. Berthold realized that something manufactured by the testes was responsible for maintenance of virilization; returning the testicles to the intra-abdominal cavity of the rooster led to a restoration of its phenotype. Because Berthold had severed the neurovascular connection of the testis to the body, he correctly surmised that the testicles produced a secretion that must act through the circulatory system. It was Charles Brown-Sequard who confirmed the testes as the source of the secretion of this "hormone." After he extracted fluid from the testis of dogs and injected himself with this fluid, he experienced improvements in strength, appetite, and mentation and published the results. The exact chemical messenger remained unidentified until 1935, when David et al. published the structure of the steroidal chemical and named the compound "Testosterone." Shortly thereafter, Butenandt and Ruzicka published near simultaneous manuscripts describing the methods for testosterone synthesis, which resulted in them subsequently sharing the Nobel Prize.

Men with low or absent testosterone production suffer significant and recognizable physical consequences. Classic male hypogonadism, a relatively rare condition, may be due to central (hypogonadotropic hypogonadism) or peripheral (hypergonadotropic hypogonadism) loss of testosterone production. German, Swiss, and Dutch chemists successfully synthesized synthetic testosterone in the 1930s, and an oral form of testosterone for replacement therapy became available in the 1940s. Testosterone replacement therapy (TRT), which maintains serum testosterone levels in the normal range, rapidly became the standard of care for hypogonadal men due to

C.F. Donatucci, MD (✉)
Department of Urology, Duke University, 5 Trafalgar Place, Durham, NC 27707, USA
e-mail: donat001@mc.duke.edu

W.J.G. Hellstrom (ed.), *Androgen Deficiency and Testosterone Replacement: Current Controversies and Strategies*, Current Clinical Urology,
DOI 10.1007/978-1-62703-179-0_1,

the availability of synthetic testosterone. The initial indication for TRT was somewhat limited. It included men suffering from genetic abnormalities like Kleinfelter's syndrome, true endocrinopathies with near castrate levels of testosterone, and the most severe form of hypogonadism. The results of TRT in these men were dramatic, and the most notable was the restoration of erectile function. As a consequence, the use of testosterone as therapy for erectile dysfunction became commonplace, primarily due to the absence of any other effective therapy at the time. This "off label" use of testosterone for the treatment of erectile dysfunction was marginally effective at best. Testosterone was first approved in 1972 by the Food and Drug Administration (FDA) as a treatment for the signs and symptoms of male hypogonadism. The use of TRT in men with prostate cancer has been contra-indicated since its original approval. The contraindication for the use of TRT in hypogonadal men with a history of prostate cancer remains controversial, despite the absence of evidence of significant risk.

Medicalization of Aging

All living organisms have a finite life cycle, and all forms of life ultimately expire at the end of that cycle. Disease represents conditions of nature and environment that may shorten the life cycle of any individual. Good health allows the individual to live out the full lifespan of the species. Organized medicine represents efforts to identify and modify disease by scientific means, and to reverse the effects of disease on the life cycle. Scientists, physicians, and other health care providers study the accumulated scientific and experiential body of knowledge, and bring the results of these efforts to the benefit of the individual patient and the health of society.

Aging in men leads to decline in multiple physiologic and psychological parameters. The decline is progressive and is in part due to decreasing androgen production. The reduction of serum levels of total, free, and bioavailable testosterone throughout aging has been well documented. The symptoms attendant to this drop in androgens, like depression, loss of energy, and sexual dysfunction, are so common that some would say that they are not symptoms at all but the normal consequence of age itself. The physical effects of aging broadly recognized include weight gain, loss of muscle mass, and skin changes. Menopause is a universal event for aging women; however, the treatment remains controversial. "Male menopause" or "andropause" are terms associated with a male syndrome similar to menopause, but sufficiently different. This causes some to question the wisdom of androgen replacement in men suffering a significant decrease in quality of life. Efforts to reverse the effects of aging are themselves timeless (e.g., the search for the "fountain of youth"). The role of the testes in rejuvenation was first appreciated by Eugen Steinbach who, together with urologist Robert Lichtenstern, developed a surgical procedure a century ago (vasectomy) that purportedly led to the improvement in the symptoms of aging [2].

Initial testosterone preparations were administered by oral or parenteral routes; while effective in raising serum testosterone levels, adverse effects in some men

Table 1.1 Ovid search of published manuscripts 2003–2012

Publications and clinical trials 2003–2011	Clinical trial	Series	Review	Other[a]
Testosterone and sexual dysfunction	15	38	46	69
Testosterone and frailty	9	4	27	2
Testosterone and well-being	5	6	35	16
Testosterone and cognitive dysfunction	0	0	3	0
Testosterone and disability	4	1	12	9
Testosterone and weakness	5	1	18	0
Testosterone and vitality	3	4	7	3

[a]Articles concerning testosterone and function of interest but not related to aging male

accompanied each. The development of transdermal testosterone delivery by skin patch, approved by the FDA in 1995, led to an increase in disease awareness and testosterone use. Concurrently, the population of patients and the indications for testosterone replacement were broadening. Since the introduction of transdermal testosterone gel in 2000, the market for TRT has increased dramatically, for men with age related decline of testosterone. It was also promoted as a reversal for the effects of aging in men. In 2002, this led the National Institute of Aging and the National Cancer Institute to request that the Institute of Medicine (IOM) develop "an independent assessment of clinical research on testosterone therapy and make recommendations for future direction for this field." The resulting white paper was published in 2004 and suggested that small efficacy trials be performed in seven clinical areas prior to undertaking a large and long-term study of safety in testosterone replacement in the aging male [3]. The IOM report recommended studies in older men with low testosterone (recommendation 1); with specific requests for short-term randomized double-blind, placebo-controlled efficacy trials (recommendation 2). The IOM panel felt that long-term risks and benefits should be determined by further clinical trial only if substantial evidence of clinically significant benefit was witnessed in the short-term studies (recommendation 3).

To gauge how successful the IOM recommendations have been to date, an OVID search was performed on the seven keywords (strength/frailty/disability, cognitive function, sexual function, and vitality/well-being/quality of life) and cross-referenced with the keyword "testosterone." All publications published between 2003 and 2011 sharing testosterone and one of the keywords from 2004 to 2012 were identified, characterized (clinical trial, series, review, and other), and tabulated (Table 1.1). The list is not exclusive and a publication with more than one keyword identified may appear more than once. As with many clinical conditions, when confirmatory evidence of treatment efficacy is lacking, the number of review articles often exceeds the number of clinical trial manuscripts. When http://www.clinicaltrials.gov web site was searched for testosterone clinical trials, a total of 598 clinical trials were listed. After elimination of trials of testosterone in women, prostate cancer, nongeriatric related male trials, and pharmacokinetic trials, the remaining trials were plotted based on the date the trial was first received (Fig. 1.1). An initial increase in recorded trials occurred in the first few years after the IOM publication, which has now receded somewhat.

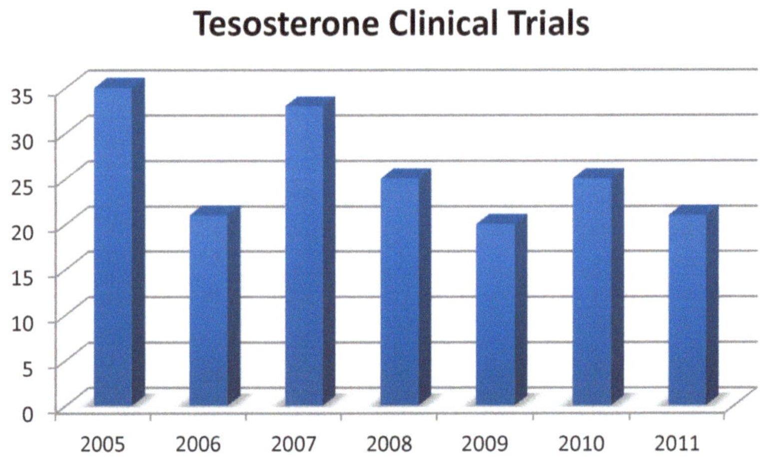

Fig. 1.1 Number of testosterone clinical trials by year of first registration 2005–2011 (based on data from http://www.ClinicalTrials.Gov)

Sexual Dysfunction

Testosterone is closely tied with sexual function, so the majority of publications in the immediate period after the publication of the IOM report dealt with the intersection of testosterone and sexual function. Several manuscripts investigated the validity of new patient reported survey instruments in aging men to see if they were truly reflective of the association between aging and androgens. Others continued to investigate the "aging versus disease state" conundrum: whether sexual dysfunction is a result of abnormally low testosterone levels or just another physical aspect of aging. Finally, a large meta-analysis of 17 studies published between 1979 and 2003/2004 outlined the positive and negative effects of TRT on sexual function.

In a study of two commonly used patient reported outcome (PRO) instruments, Basar et al. investigated the relationship between testosterone levels and symptoms using the Aging Male Symptoms (AMS) score and International Index of Erectile Function (IIEF) [4]. The AMS and IIEF questionnaires were administered to 348 men (21–67 years; mean 49.6 years); serum sex hormone levels (total testosterone (TT), free testosterone (FT), estradiol (E2), and dehydroepiandrosterone-sulfate (DHEA-S)) were drawn. Serum DHEA-S and FT levels and age correlated significantly with the IIEF scores; although serum total testosterone, FT, and DHEA-S levels correlated significantly with the andrologic domain of AMS, the only correlation of the total AMS score was with age.

The Androgen Deficiency of Aging Men (ADAM) questionnaire, another popular PRO, was also evaluated for efficacy in diagnosing hypogonadism. Blumel et al. administered the ADAM questionnaire and drew serum sex hormone profiles on 96 men aged 40 years and older in a cross-sectional study in Chile [5]. Total testosterone, sex hormone binding globulin (SHBG), and albumin were measured, and bioavailable testosterone was calculated. Standard scoring for the ADAM questionnaire was used to evaluate for androgen deficiency (if items 1 or 7 or any three other questions of the ADAM questionnaire were positive). A total of 78 men (81.3%) were identified as androgen deficient by the ADAM questionnaire; however, available testosterone confirmed the diagnosis in only 27 cases (28.1%). Low libido was a better predictor of hypogonadism by itself, rather than the ADAM questionnaire (63.3% sensitivity and 66.7% specificity). For now we do not have a statistically valid, rigorous PRO for androgen deficiency in aging men, and the need remains, a topic discussed in a later chapter of this text.

While efforts to find a good PRO continue, others tried to determine whether there exists a threshold at which patients become symptomatic, or whether each symptom may have an individual serum testosterone threshold at which it emerges. In such an attempt, Lackner et al. evaluated 675 healthy men using the Aging Male Symptoms (AMS) scale, the Beck Depression Index (BDI) and the International Index of Erectile Function (IIEF), and attempted to correlates scores with AM serum testosterone levels [6]. The patients were divided into two groups: those with symptoms and those without. Patients suffering from lack of concentration, decreased libido, listlessness, and both somatic/psychological symptoms by AMS demonstrated testosterone levels that were different from those in men without these symptoms. However, this association was lost when multivariate analysis was applied; loss of libido, lack of vigor, and sexual dysfunction were associated with age rather than with testosterone. Thus, levels of serum testosterone at which specific symptoms emerged could not be identified in this study.

Two manuscripts examined the relationship of testosterone levels with erection and ejaculation. Gades et al. evaluated the association between sex hormone serum levels, erectile function, and sexual drive in a population-based sample of men (the Olmsted County Study of Urinary Symptoms and Health Status) [7]. In a random sample of men residing in Olmsted County, MN, 414 men completed the Brief Male Sexual Function Inventory (BMSFI), underwent physical examination, and had serum hormone measurements performed. Of these men, 294 had a regular sexual partner and androgen measurements after 14 years of follow-up. At 14 years, total testosterone and erectile function (but not libido) demonstrated significant correlation, despite adjustment for age. Bioavailable testosterone was significantly correlated with both erectile function and libido.

Corona et al. evaluated the role of testosterone and hypogonadism in the control of the ejaculatory reflex, comparing subjects with premature ejaculation (PE) or delayed ejaculation (DE) to those without ejaculatory dysfunction [8]. Serum hormonal and biochemical parameters were studied in 2,437 men with sexual dysfunction. Premature ejaculation was reported by 714 (25.9%) and delayed ejaculation by 121 (4.4%). Testosterone levels significantly correlated with the type of ejaculation. The youngest men (25–40 years) with PE had higher TT and FT levels; the oldest (55–70 years), with DE, had lower TT and FT levels. As would be expected from these results, hypogonadism was more common in men with DE (26%) and least common in men suffering PE (12%). Adjustment for age and libido did not change the results.

Positive correlations between threshold testosterone levels and specific sexual symptoms were found in a study of TRT reported by Seftel et al. [9]. Four hundred and six aging symptomatic hypogonadal men (mean age 58 years) were randomized to transdermal testosterone gel (50 and 100 mg/day), transdermal testosterone patch, or placebo. Patients were evaluated at 30 and 90 days after initiation of treatment for significant change in the frequency of intercourse, nighttime erections, and libido. At day 30, a significant increase was seen for all three primary outcome measures for those on 100 mg/day T gel compared to the others; the results at 90 days for sexual desire and nighttime erections vs. placebo were similar. The authors defined a "threshold average daily serum T level for sexual response" which was the testosterone level at which (1) while the testosterone level might be within the normal range, the sexual response was no different from that of the group of subjects with the lowest serum T level (0–300 ng/dL); and (2) there was a significant change when treated with testosterone compared with that of the group of subjects with the lowest serum level. This threshold f was 400 ng/dL for nighttime erections; at or below this serum level, the frequency of nighttime erections was no different than that for hypogonadal men. For frequency of sexual intercourse, the "threshold average daily serum testosterone level was 500 ng/dL, and it was 600 ng/dL for sexual desire. While the authors felt these data demonstrated a clear relationship between restoring serum T concentrations and improvement in certain parameters of sexual function, they do state that the study was not designed to explicitly measure a threshold serum level.

Isidori et al. performed a systematic review and meta-analysis of placebo-controlled studies published in the past 30 year period (1975–2004) [10]. MEDLINE, the Cochrane Library, EMBASE, and Current Contents were reviewed, of which 17 randomized placebo-controlled trials were found to be eligible. The total number of evaluable patients over the 17 studies was 656, of which 284 were randomized to testosterone and 284 to placebo (P), and 88 were treated in crossover. The majority of the patients received parenteral testosterone, though a few men received transdermal or buccal testosterone. The average length of therapy across the studies was 3 months (range 1–36 months). Results of the meta-analysis revealed that in hypogonadal men with an initial testosterone level below 12 ng/dL, TRT led to an improvement in multiple parameters: nocturnal erections, sexual thoughts and motivation, number of successful intercourses, scores for erectile function, and overall sexual satisfaction. Testosterone replacement had no effect on sexual function in eugonadal men and was equivalent to placebo. The effects of T on erectile function, but not libido, were inversely related to the mean baseline T concentration. Results of this meta-analysis must be viewed critically as the majority of the studies were performed during a period when sexual function, and in particular erectile function, was the most common outcome used to judge clinical response to testosterone replacement, and the physiology of erection and pathophysiology of erectile dysfunction were poorly understood. Conclusions from studies performed during this period are not necessarily valid given our understanding of androgens and erectile function today.

Well-Being/Quality of Life/Cognitive Function

As men get older, they experience many conditions, often together, that eventually result in the inability to perform many activities of daily living. Some men may experience an increased propensity to fall and decreased independence. Elderly men also have increased incidence of anemia, higher rates of metabolic syndrome, decreased sexual function, and memory impairment. These conditions likely have multiple causes, but one likely contributing cause is a low serum testosterone concentration. When young hypogonadal men are treated with testosterone, they experience improvements in sexual function, muscle mass and strength, bone mineral density (BMD), and sense of well-being. In aging men, the benefits of testosterone therapy are controversial, particularly those related to psychosocial improvement. A limited number of studies have examined these issues since the IOM report, the following are the most important.

The degree to which androgen deficiency leads to a decline in the quality of life in elderly is unknown. To examine this question, Finas et al. investigated 24 hypogonadal men aged 50 years or older (defined as free testosterone <200 ng/dL) and compared them to 24 eugonadal age-matched controls. All of the men were under treatment for benign prostatic hyperplasia. The Short Form (SF)-12 Health Survey (physical health and mental health index) assessed health-related "Quality of Life." The investigators modified the questionnaire to broaden it, importing "vitality" and "psychological well-being" scales from SF-36. The SF-12 physical health index was reduced in the hypogonadal group, but the mental health index was not. Patients with low FT demonstrated lower scores on vitality than eugonadal men, but no differences were detected in psychological well-being between hypogonadal and eugonadal men [11].

T'Sjoen et al. used the AMS rating scale to assess the relationship of male aging symptoms and androgen levels in asymptomatic elderly men in generally good health [12]. After the AMS questionnaire was completed, serum sex steroids levels, sex hormone binding globulin, gonadotropins, and physiologic responses were measured in 161 ambulatory, elderly men, aged 74–89. Mild psychological and mild to moderate somato-vegetative symptoms were associated with diminished serum testosterone levels as assessed by AMS. However, none of the three AMS domain scale scores significantly correlated with testosterone, free testosterone or bioavailable testosterone. In healthy elderly men, the AMS was not useful in predicting androgen levels because no consistent relationship was demonstrated.

Androgens may play a role in maintenance of cognitive function in men. Patients suffering from derangements of cognitive function such as Alzheimer's disease may benefit from TRT. Previously, Hogervorst et al. found lower levels of testosterone in men with Alzheimer's disease compared with controls [13]. To determine whether abnormal pituitary regulation of androgens was responsible for the testosterone deficiency, the investigators compared sex hormones (follicle stimulating hormone (FSH) and luteinizing hormone (LH), and sex hormone binding globulin (SHBG)) and testosterone in 45 men with Alzheimer's, 15 men with other types of dementia, and 133 elderly controls. There were no observed differences in LH, FSH, or SHBG levels between Alzheimer's disease patients and controls. However, testosterone levels were significantly lower in the men with Alzheimer's.

The association of depression and low testosterone in men is recognized; what is not as well known is whether there is a causal relationship between low testosterone and depression. In a large cross-sectional study, Almeida et al. sought to determine whether the association between serum testosterone concentration and mood in older men is independent of physical comorbidity [14]. The authors investigated 3,987 men aged 71–89 years from a community-based setting. Patients were evaluated using the Geriatric Depression Scale (GDS-15) to assess mood; physical health was assessed with Physical Component Summary score of the 36-Item Short Form Health Survey. Two hundred and three men (5.1%) had depression. These men had significantly lower total and free testosterone concentrations than nondepressed men. Depressed men also had higher obesity levels and more physical limitations. After adjusting for these factors and for age, the association between depression and low total and free testosterone concentrations did not change.

Giltay et al. examined the reversibility of hypogonadal induced depression secondary to testosterone deficiency in a randomized, placebo-controlled, double-blind, phase III trial (ClinicalTrials.gov identifier: NCT00696748) [15]. In this trial, 184 men with metabolic syndrome and low testosterone were randomized 2:1 to receive testosterone or placebo injections. Mood, well-being, and sexual function were assessed at three time periods (baseline, 18 and 30 weeks) using the Beck Depression Inventory (BDI-IA), Aging Males' Symptoms (AMS) scale, and International Index of Erectile Function 5-item (IIEF-5) scale. Restoration to a eugonadal state led to significant improvements in all three parameters (BDI-IA, AMS, and IIEF); the greatest improvement occurred in men with a baseline total testosterone level <7.7 ng/dL.

Disability/Frailty/Vitality

The majority of studies published since the IOM report have dealt with the impact of androgen deficiency and replacement on frailty in the aging, and the disability and loss of vitality that accompany frailty. These reports are comprehensive and come from large populations worldwide. The outcomes are tangible and more easily measured (i.e., bone mineral density, muscle strength, etc.). There is also more certainty that therapeutic interventions are causal to the observed improvements (Table 1.2).

Schaap et al. reported on a cross-sectional population-based study, based on data from the Longitudinal Ageing Study Amsterdam (LASA). Sex hormones were measured, and physical performance, functional limitations, and muscle strength were assessed. Analysis of falls was performed prospectively for 3 years [16]. Men in the highest quartile of the estradiol/SHBG ratio had significantly higher physical performance scores than men in the lowest quartile. Serum levels of total testosterone

Table 1.2 Therapeutic effects of testosterone replacement on physical measures of disability/frailty/vitality

	Number of patients	TRT method/duration	Assessment
Kenny et al. [22]	131	TD × 24 months	BMD(±), BC(+), SH(+), strength(−), physical performance(−)
Srinivas-Shankar et al. [23]	274	Transdermal × 6 months	MS(+), BC(+), AMS (somatic+, psychologic−, sexual+)
Wittert et al. [25]	76	Oral TU × 12 months	BC(+), MS(+), SH(+), AE(−)
Permpongkosol et al. [24]	161	Depo TU mean 90 weeks	SH(+), BMI(−), WC(+), %BF(−), AMS(−)

BMD bone mineral density; *BC* body composition; *SH* sex hormones; *MS* muscle strength; *AE* adverse effects; *WC* waist circumference; *BF* body fat

were positively associated with muscle strength. Calculated bioavailable testosterone levels were positively associated with physical performance and muscle strength. Low levels of sex hormones were associated with impaired mobility and low muscle strength in men.

In an observational study of 1,705 men in Australia, Travison et al. used longitudinal measurements to assess sex hormones and their relationship to the prevalence and progression of frailty in older men [17]. Measurements were obtained at baseline (2005–2007) and at 2-year follow-up (2007–2009). Frailty syndrome was measured according to the Cardiovascular Health Study (CHS) and Study of Osteoporotic Fractures (SOF) indices. Significant age-adjusted associations were seen between serum androgens and estrogens and concurrent frailty. Subjects in the lowest T quintile had 2.2-fold odds of exhibiting greater CHS frailty as compared with the highest T quintile ($P<0.001$). A 2-year decrease of one standard deviation in T, calculated free T, or LH was associated with a 1.2- to 1.3-fold increase in the odds of progression (increase in severity) of frailty. Control for comorbid medical conditions did not affect the age-related changes. The author concluded that an Serum androgens and estrogens may contribute to the development or progression of frailty in men.

Cawthon et al. reported a large multicentered cross-sectional/longitudinal study of sex hormones and frailty based in the United States [18]. The Osteoporotic Fractures in Men (MrOS) study involved 1,469 men at least 65 year old; 1,245 men had frailty status reassessed after 4 years. Frailty was assessed by using a modification of the CHS, with patients exhibiting three or more symptoms (shrinking/sarcopenia, weakness, slowness, low activity level, and exhaustion). Bioavailable testosterone was the only serum sex hormone associated with frailty. Men in the lowest quarter of bioavailable testosterone had 1.39-fold increase in odds of greater frailty status compared to men in the highest quartile, and a 1.51-fold increase in odds of greater frailty status 4 years later. Comorbid conditions had little effect. A second US based study used a modification of the Cardiovascular Health System to assess sex hormones and frailty in older men [19]. The Massachusetts Male Aging Study Examined 646 men aged 50+ with frailty defined as the presence of three or more of the following: weight loss, exhaustion, low physical activity, slowness, and weakness. TT, free T, and SHBG levels were investigated for association with frailty and with degree of frailty. Total and free T were not associated with frailty in these aging men; however, SHBG did show a significant association with frailty. Grip strength and physical activity (but not exhaustion, slow walking, or weight loss) were associated with total T levels, whereas SHBG was related to weight loss, exhaustion, and physical activity.

Tajar et al. [20] reported the results of the European Male Aging Study. Three thousand two hundred and nineteen men had frailty assessed as an index (FI) according to the number (out of 43 possible) of health deficits (symptoms, signs, and functional impairments), and relationships between FI and hormone levels (as outcomes). Results were explored using regression models. Mean FI was highest in the oldest group and higher levels of FI were significantly associated with lower levels of total T, free T, and DHEA-S and higher levels of gonadotropins and SHBG. In the "Health In Men" study of frailty and androgens, Hyde et al. examined 3,616 elderly men living in Australia [21]. Sex hormones (testosterone SHBG, LH, and calculated free testosterone) were measured. The authors used a "FRAIL scale"; patients were scored on five physical tasks: fatigue, difficulty climbing a flight of stairs, difficulty walking more than 100 m, more than five illnesses present, or weight loss greater than 5%. Men with three or more positive symptoms were deemed frail. Frailty was assessed at baseline; of the 3,616 men, 548 (15.2%) were judged frail (at least three deficits). At following reassessment in 2008–2009, frailty increased to 23.0% (364/1,586). Of the sex hormones evaluated, only free testosterone predicted frailty after statistical adjustment for age and other comorbid factors.

Four clinical trials were published in the period after the IOM report, which document the beneficial effects of androgen replacement in frail men. Though much smaller in number, the aging patients demonstrated measurable improvements in objectively measurable physiologic parameters. Three different routes of administration were used to achieve normal serum testosterone levels (oral, parenteral, and transdermal). In a single center randomized study, Kenny et al. evaluated the effects of transdermal testosterone on bone and muscle in older men with low bioavailable testosterone levels [22]. This study was designed after publication of the IOM report and incorporated the IOM recommendations for small efficacy studies: focus on health outcomes for which preliminary evidence of efficacy and limited alternative therapies exist. One hundred and thirty-one aging men (mean age 77 years) with low testosterone, history of fracture, or osteoporosis (*T*-score less than −2.0) were included in the study. Frailty was assessed based on Fried and colleagues' parameters (frail = 3–5 characteristics, intermediate frail = 1–2 characteristics, nonfrail ($n=0$ characteristics). Patients received 5 mg/day of testosterone or placebo for 12–24 months. Additionally, all men received calcium (1,500 mg/day diet and supplement) and cholecalciferol (1,000 IU/day). Multiple outcome measures were assessed, including BMD (hip, lumbar spine, and mid-radius), body composition, sex hormones, calcium-regulating hormones, bone turnover markers, strength, physical performance, and safety parameters. Ninety-nine men (75.6%) completed 12 months. Sixty-two (47.3%) completed end of therapy (mean 23 months). BMD on testosterone increased 1.4% at the femoral neck and 3.2% at the lumbar spine. There was an increase in lean mass and a decrease in fat mass in the testosterone group but no differences in strength or physical performance. There were no differences in safety parameters.

Table 1.3 Observational trials of testosterone and measures of disability/frailty/and vitality

	Number of patients	Age (years)	Outcome measure
LASA [16]	623	68–88	ED(−), MS(+), PP(+), FL(+)
CHAMP [17]	1,645	>69	Frailty Index (CHS(+), FOS(+))
MrOS [18]	1,245	>65	Frailty status(−), T(−), BioT(+)
EMAS [20]	2,219	40–79	T(+)
MMAS [19]	646	50–83	SH, Frailty Index
HIM [21]	3,616	76–93	FRAIL Scale, SH

ED erectile dysfunction; *MS* muscle strength; *PP* physical performance; *FL* falls; *SH* sex hormones

Transdermal testosterone gel was also used by Srinivas-Shankar et al. to investigate testosterone replacement on muscle strength, physical function, body composition, and quality of life in intermediate-frail and frail elderly men [23]. In a single-center study, 274 elderly men at least 65 years of age with hypogonadism (total T ≤ 12 nmol/L or free T at or below 250 pmol/L) were randomized to transdermal T (50 mg/day) or placebo gel for 6 months. Outcome measures included muscle strength, lean and fat mass, physical function, and self-reported quality of life. At the end of the study, muscle strength improved in the T group (vs. placebo at 6 months), while lean body mass increased and fat mass decreased significantly. Somatic and sexual symptom scores decreased with T treatment.

In Thailand, Permpongkosol et al. conducted a retrospective review of 161 men with symptomatic late onset hypogonadism who had received treatment with parenteral testosterone undecanoate. The effects of the therapy on body composition, lipids, and psychosexual complaints were noted [24]. Late onset hypogonadism was defined as pretreatment T < 300 ng/day. The mean duration of treatment was 90.6 weeks. One hundred men had used parenteral TU for >12 months. Body mass index (BMI), waist circumference, percentage body fat, total cholesterol, high-density lipoprotein (HDL) cholesterol, low-density lipoprotein (LDL) cholesterol, triglycerides, prostate-specific antigen (PSA), and hematocrit were measured. Further, the AMS scale and the IIEF (IIEF-5 and 15) were scored. Testosterone replacement led to a significant decline in waist circumference and percentage body fat, but no change in BMI. The psychological, somato-vegetative, and sexual factor scores of subscales of AMS decreased. Erectile function domain, orgasmic function domain, sexual desire domain, intercourse satisfaction domain, and overall satisfaction domain improved. No clinically significant adverse effects occurred in these men. Oral testosterone supplementation was used in an attempt to prevent muscle loss in eugonadal frail men over a 12-month period [25]. Seventy-six men, aged 60 years or older, received testosterone undecanoate (80 mg twice daily) or placebo for 1 year. Measurements of body composition, muscle strength, hormones, and safety parameters were obtained at 0, 6, and 12 months. Lean body mass increased and fat mass decreased in the testosterone treated group. There were no significant effects on muscle strength. Again, no clinically significant adverse effects were seen.

Over 10,000 aging men from four continents were included in the studies listed above (Table 1.3). All of the studies related increasing frailty status to decreasing androgens. However, there was disagreement between studies as to which androgen measurement was predictive of frailty. Each study attempts to objectively assess frailty status. Many use commonly accepted scales (Cardiovascular Health System, and Study of Osteoporotic Fractures), but modifications were made locally making comparisons and measurements sometimes difficult to interpret. Methodologies, though similar, do not utilize the IOM recommended trial design. The prospective treatment studies were composed of different testosterone preparations and duration and measured different outcomes. All of these factors reduce the ability to generalize the conclusions of individual studies.

One study, published in the period since the IOM report, has received significant notoriety. The Testosterone In Older Men With Mobility Limitations (TOM) trial was a single center, placebo-controlled, randomized clinical trial designed to comprehensively determine the effects of testosterone administration on muscle strength and physical function in older men with mobility limitations [26]. Community-dwelling men, 65 years of age or older, with limitations in mobility and a total serum testosterone level of 100–350 ng/dL or a free serum testosterone level <50 pg/mL, underwent randomization to placebo gel or testosterone gel for 6 months. Adverse events were categorized with the use of the Medical Dictionary for Regulatory Activities (MedRA). At midterm review, the data and safety monitoring board recommended that the trial be discontinued early. A significantly higher rate of adverse cardiovascular events was found in the testosterone group compared to the placebo group. A total of 209 men (mean age, 74 years) were enrolled at study termination. Of the 209 men, a total of 23 subjects in the testosterone group, as compared to 5 in the placebo group, had cardiovascular-related adverse events. The relative risk of a cardiovascular-related adverse events remained constant throughout the 6-month treatment period. The authors did admit in the manuscript that the small size of the trial and the unique population prevented broader inferences from being made about the safety of testosterone therapy (ClinicalTrials.gov number, NCT00240981).

The TOM trial raised concerns about the safety of testosterone therapy in aging men, but was justifiably criticized for several shortcomings. The initial dose of testosterone used in the study was greater than the recommended starting dose in the product label. Patients were also titrated to doses above the maximal approved dose. Several of the adverse effects experienced in the treatment group were known effects of testosterone replacement (peripheral edema) and clinically insignificant in nature. Yet, appropriate concern was used when the Data Safety Monitoring committee made the decision to terminate the study early. Definitive demonstration of an acceptable risk/benefit ratio, as defined in the IOM report, remains unfulfilled.

The Testosterone Trial in Older Men

Officially known as "Randomized, Placebo-controlled, Double-blind Study of Five Coordinated Testosterone Treatment Trials in Older Men," this multicentered, randomized, placebo-controlled, double-blind treatment efficacy and safety study meets the challenge for clinical trial design as recommended by the IOM 2004 report [27]. The testosterone "Trial" will determine if 1 year of active testosterone replacement will lead to improvement in five primary outcome measures: walking speed, sexual activity, vitality scale, verbal memory test, and correction of anemia (ClinicalTrials.gov Identifier: NCT00799617). The trial is sponsored and funded by the University of Pennsylvania with the National Institute for Aging, the National Institute of Neurological Disorders and Stroke, the Eunice Kennedy Shriver National Institute of Child Health and Human Development and the National Heart Lung & Blood Institutes serving as collaborators, with additional assistance from Abbott Corporation. Eight hundred men, aged 65 or older, with related androgen deficiency will enroll at 1 of the 12 participating medical centers beginning in November 2009 and ending in June 2015. The laboratory definition of low testosterone in this study is a total morning (drawn between 7 and 10 a.m.) serum concentration of <275 and <300 ng/dL at each of two screening visits. Within the "Testosterone in Men" trial, two sub-trials will be performed prospectively. Two additional trials have been incorporated into the T Trial. One, a cardiovascular trial, examines the effects of testosterone treatment on cardiovascular risk factors compared to placebo. In the second, the bone trial attempts to document an increase in volumetric trabecular BMD of the lumbar spine.

The testosterone trial is the result of multiple investigators' efforts to incorporate the IOM recommendations into a large safety and efficacy trial. There are commonly agreed upon evaluative tools, PRO measures, primary and secondary physical outcomes, and adverse event reports. The large size of the trial requires support of both government and industry. We will await the results of the testosterone trial with great interest.

Disclaimer The opinions expressed by the author are his alone and do not reflect those of Eli Lilly & Company.

References

1. Freeman ER, Bloom DA, McGuire EJ. A brief history of testosterone. J Urol. 2001;165(2): 371–3.
2. Schultheiss D, Denil J, Jonas U. Rejuvenation in the early 20th century. Andrologia. 1997;29(6):351–5.
3. Liverman CT, Blazer DG, editors. Testosterone and aging: clinical research directions. Washington, DC: National Academies Press; 2004.
4. Basar MM, Aydin G, Mert HC, et al. Relationship between serum sex steroids and Aging Male Symptoms score and International Index of Erectile Function. Urology. 2005;66(3):597–601.
5. Blumel JE, Chedraui P, Gili SA, Navarro A, Valenzuela K, Vallejo S. Is the Androgen Deficiency of Aging Men (ADAM) questionnaire useful for the screening of partial androgenic deficiency of aging men? Maturitas. 2009;63(4):365–8.
6. Lackner JE, Rucklinger E, Schatzl G, Lunglmayr G, Kratzik CW. Are there symptom-specific testosterone thresholds in aging men? BJU Int. 2011;108(8):1310–5.
7. Gades NM, Jacobson DJ, McGree ME, et al. The associations between serum sex hormones, erectile function, and sex drive: the Olmsted County Study of Urinary Symptoms and Health Status among men. J Sex Med. 2008;5(9):2209–20.

8. Corona G, Jannini EA, Mannucci E, et al. Different testosterone levels are associated with ejaculatory dysfunction. J Sex Med. 2008;5(8):1991–8.
9. Seftel AD, Mack RJ, Secrest AR, Smith TM. Restorative increases in serum testosterone levels are significantly correlated to improvements in sexual functioning. J Androl. 2004;25(6):963–72.
10. Isidori AM, Giannetta E, Gianfrilli D, et al. Effects of testosterone on sexual function in men: results of a meta-analysis. Clin Endocrinol. 2005;63(4):381–94.
11. Finas D, Bals-Pratsch M, Sandmann J, et al. Quality of life in elderly men with androgen deficiency. Andrologia. 2006;38(2):48–53.
12. T'Sjoen G, Goemaere S, De Meyere M, Kaufman JM. Perception of males' aging symptoms, health and well-being in elderly community-dwelling men is not related to circulating androgen levels. Psychoneuroendocrinology. 2004;29(2):201–14.
13. Hogervorst E, Combrinck M, Smith AD. Testosterone and gonadotropin levels in men with dementia. Neuro Endocrinol Lett. 2003;24(3–4):203–8.
14. Almeida OP, Yeap BB, Hankey GJ, Jamrozik K, Flicker L. Low free testosterone concentration as a potentially treatable cause of depressive symptoms in older men. Arch Gen Psychiatry. 2008;65(3):283–9.
15. Giltay EJ, Tishova YA, Mskhalaya GJ, Gooren LJG, Saad F, Kalinchenko SY. Effects of testosterone supplementation on depressive symptoms and sexual dysfunction in hypogonadal men with the metabolic syndrome. J Sex Med. 2010;7(7):2572–82.
16. Schaap LA, Pluijm SMF, Smit JH, et al. The association of sex hormone levels with poor mobility, low muscle strength and incidence of falls among older men and women. Clin Endocrinol. 2005;63(2):152–60.
17. Travison TG, Nguyen A-H, Naganathan V, et al. Changes in reproductive hormone concentrations predict the prevalence and progression of the frailty syndrome in older men: the concord health and ageing in men project. J Clin Endocrinol Metabol. 2011;96(8):2464–74.
18. Cawthon PM, Ensrud KE, Laughlin GA, et al. Sex hormones and frailty in older men: the osteoporotic fractures in men (MrOS) study. J Clin Endocrinol Metabol. 2009;94(10):3806–15.
19. Mohr BA, Bhasin S, Kupelian V, Araujo AB, O'Donnell AB, McKinlay JB. Testosterone, sex hormone-binding globulin, and frailty in older men. J Am Geriatr Soc. 2007;55(4):548–55.
20. Tajar A, O'Connell MDL, Mitnitski AB, et al. Frailty in relation to variations in hormone levels of the hypothalamic-pituitary-testicular axis in older men: results from the European male aging study. J Am Geriatr Soc. 2011;59(5):814–21.
21. Hyde Z, Flicker L, Almeida OP, et al. Low free testosterone predicts frailty in older men: the health in men study. J Clin Endocrinol Metabol. 2010;95(7):3165–72.
22. Kenny AM, Kleppinger A, Annis K, et al. Effects of transdermal testosterone on bone and muscle in older men with low bioavailable testosterone levels, low bone mass, and physical frailty. J Am Geriatr Soc. 2010;58(6):1134–43.
23. Srinivas-Shankar U, Roberts SA, Connolly MJ, et al. Effects of testosterone on muscle strength, physical function, body composition, and quality of life in intermediate-frail and frail elderly men: a randomized, double-blind, placebo-controlled study. J Clin Endocrinol Metabol. 2010;95(2):639–50.
24. Permpongkosol S, Tantirangsee N, Ratana-olarn K. Treatment of 161 men with symptomatic late onset hypogonadism with long-acting parenteral testosterone undecanoate: effects on body composition, lipids, and psychosexual complaints. J Sex Med. 2010;7(11):3765–74.
25. Wittert GA, Chapman IM, Haren MT, Mackintosh S, Coates P, Morley JE. Oral testosterone supplementation increases muscle and decreases fat mass in healthy elderly males with low-normal gonadal status. J Gerontol A Biol Sci Med Sci. 2003;58(7):618–25.
26. Basaria S, Coviello AD, Travison TG, et al. Adverse events associated with testosterone administration. N Engl J Med. 2010;363(2):109–22.
27. Cook NL, Romashkan S. Why do we need a trial on the effects of testosterone therapy in older men? Clin Pharmacol Ther. 2011;89(1):29–31.

Chapter 2
The Laboratory Diagnosis of Testosterone Deficiency

Akanksha Mehta and Darius A. Paduch

Introduction

The laboratory diagnosis of testosterone deficiency is challenging. Serum testosterone levels are subject to temporal variation—diurnal, seasonal, and age-related. Illness and certain medications, such as opiates and glucocorticoids, can temporarily affect testosterone concentrations through central and peripheral effects. Total testosterone concentrations are affected by alterations in sex-hormone binding globulin (SHBG), which in turn can occur for a variety of reasons, including age, medications, and medical comorbidities. There are several different assays for measurement of serum testosterone levels. Performance characteristics, linearity, reproducibility, low level limits of detection, and pre-analytical requirements vary among the assay platforms, leading to a wide variety of normal ranges reported by different laboratories. Lastly, testosterone circulates in the blood primarily bound specifically to SHBG or non-specifically to albumin or cortisol, with only 2–3% of total testosterone being free. Which of these derived testosterone measures most closely correlates with symptomatic androgen deficiency is a matter of debate. There is no currently available test that reflects tissue androgen activity. It appears that tissue responses differ across different organ systems, and change in parallel over time. Thus, androgen deficiency may become apparent at different ages within an individual or a population. With the multitude of factors affecting the laboratory evaluation and interpretation of testosterone levels, it is no surprise that a universally accepted

A. Mehta, MD
Department of Urology, Weill-Cornell Medical College, New York, NY, USA

D.A. Paduch, MD, Ph.D (✉)
Department of Urology and Reproductive Medicine, Weill-Cornell Medical College, 525 East 68th Street, New York, NY 10065, USA

Consulting Research Services, Inc., Red Bank, NJ, USA
e-mail: darius.paduch@mac.com

W.J.G. Hellstrom (ed.), *Androgen Deficiency and Testosterone Replacement: Current Controversies and Strategies*, Current Clinical Urology,
DOI 10.1007/978-1-62703-179-0_2,

definition of testosterone deficiency is lacking. Under the leadership of the Centers for Disease Control, there is now a national effort to address this problem [1].

In this chapter, we discuss the currently used assays for testosterone measurement, their utility and limitations, and the implications for clinical practice.

Defining Testosterone Deficiency

There is no consensus among endocrinologists, urologists, and clinical pathologists as to what defines a "low" testosterone level. Published norms for serum testosterone are mostly based on studies that were not specifically designed to establish norms in men with normal sexual and reproductive function. Recently published guidelines from the Endocrine Society maintain that total testosterone levels below 300 ng/dL are diagnostic of hypogonadism, and higher levels are normal [2]. Meanwhile, a consensus statement from the ISA,[1] ISSAM,[2] EAU,[3] EAA,[4] and ASA[5] recommends that total testosterone levels above 350 ng/dL do not require treatment, while levels below 230 ng/dL do [3]. For levels between 230 and 350 ng/dL, the recommendation is to repeat the total testosterone with SHBG for calculation of free testosterone (free T), or direct measurement of free T by equilibrium dialysis (Level 2a evidence, Grade A recommendation) [3]. Similarly, we have previously recommended that men with total testosterone <200 ng/dL be treated as hypogonadal, those with total testosterone >400 ng/dL be considered normal, and those with total testosterone 200–400 ng/dL be treated based on clinical presentation [4].

Some authors have advocated for the use of free or bioavailable testosterone (free T, bioavailable T) to aid in biochemical diagnosis of hypogonadism, especially when results of the total testosterone assay are equivocal or fail to reflect the clinical presentation. Though there are no generally accepted lower limits of normal free testosterone for the diagnosis of hypogonadism, according to expert opinion, a free testosterone level below 65 pg/mL can provide supportive evidence for treatment (Level 3 evidence, Grade C recommendation) [3]. Corresponding values for bioavailable testosterone depend on the method used, and are not generally available [5].

Given the high degree of variability among different assays, and a wide range of testosterone levels in a variety of patients of different ages and ethnicities, it is no surprise that attempts to establish a uniform threshold that accurately distinguishes hypogonadal and eugonadal men have been unsuccessful, and such a threshold, therefore, remains quite arbitrary. Considering all this, it must be

[1] ISA: International Society of Andrology.

[2] ISSAM: International Society for Study of the Aging Male.

[3] EAU: European Association of Urology.

[4] EAA: European Association of Andrology.

[5] ASA: American Society of Andrology.

emphasized that in addition to laboratory values, the diagnosis of testosterone deficiency must take into account the presence of clinical signs and symptoms suggestive of hypogonadism.

Prevalence of Testosterone Deficiency

In a multiethnic, population-based observational study of 1,475 men aged 30–79 years, Araujo et al. observed the prevalence of symptomatic androgen deficiency to be 5.6% (95% CI, 3.6–8.6%). The prevalence was lower in men less than 70 years old (3.1–7.0%), but increased substantially with age (18.4%) [6]. Longitudinal population-based studies of aging men have also demonstrated that both total and free serum testosterone decline with age, with a concomitant increase in SHBG levels [7–10]. It is estimated that by the year 2030, approximately 6.5 million American men aged 30–80 years of age may be diagnosed with androgen deficiency [6]. The importance of accurate and reliable laboratory testing in establishing the diagnosis of androgen deficiency is evident, with obvious implications for clinical management.

Variability in Testosterone Concentrations

Serum testosterone in men shows a wide range of variation, owing to episodic secretion, diurnal variation, week-to-week variation, and even seasonal changes [11]. The amplitude and acrophase of the diurnal variations in free and bioavailable testosterone levels are similar to that of total testosterone [12, 13]. One report has described a circadian pattern of SHBG levels as well, with peaks occurring in the early afternoon [13]. The finding has not been corroborated by other studies.

Serum testosterone levels peak in the early morning, followed by a progressive decline over the course of the day, until they reaching their nadir in the evening hours. Nadir values are approximately 15% lower than morning values, and may vary by as much as 50% [14]. Therefore, sampling time is an important consideration when measuring serum testosterone, and samples should be always obtained in the morning, between 0700 and 1100 hours [15]. This diurnal pattern is blunted in older men [12, 16]. It has been argued that morning testosterone measurements are not necessary in older men due to blunting of the circadian rhythm. However, a substantial fraction of older men, aged 65–80 years, who have low serum testosterone in the afternoon, will have normal testosterone concentrations in the morning [17].

Brambilla et al. have showed that great intra-individual variation in testosterone levels exists when samples are collected from the same individual at the same time of the day over several days [17]. Approximately 15% of healthy men may have a testosterone level below the normal range in a 24-h period. Furthermore, among men with an initial testosterone concentration in the mildly hypogonadal range, approximately 30% will have a normal testosterone on repeat measurement [17].

Day-to-day variations in testosterone concentrations may be large enough to render a single testosterone measurement inadequate to accurately characterize an individual's levels. At least two separate measurements of testosterone are needed to confidently diagnose hypogonadism.

It is known that testosterone measurements are prone to pre-analytical conditions and processing methods [18]. These include various technical factors, such as types of collection tubes used to obtain samples, sample handling, and processing. For example, storage of serum or plasma in collection tubes following centrifugation can lead to a decrease in measured testosterone. Storage in EDTA can adversely affect SHBG measurement, thus affecting the measurement of free hormones [18]. Pre-analytical factors can affect the variability in diurnal levels of testosterone as well. Samples obtained in the morning are typically transported and processed the same day, but samples drawn in the afternoon may be stored for processing on the following day, which may introduce analytical errors.

Circulating testosterone levels are influenced by a variety of conditions, including medications, acute illness, sexual activity, and SHBG concentrations. All must be taken into consideration when ordering or interpreting any testosterone assay [18, 19]. The suppression of testosterone is particularly profound in men on methadone maintenance therapy because of its long duration of action. Acute illness can also significantly lower serum testosterone, albeit temporarily. The Endocrine Society guidelines recommend that the diagnosis of androgen deficiency should not be made during an acute illness [2].

Testosterone: Total, Free, and Bioavailable

Testosterone circulates in the body unbound or bound to either SHBG, albumin, or Corticosteriod-binding globulin(CG) [20] (Fig. 2.1). SHBG-bound testosterone represents approximately 44% of the total T, is tightly bound, and therefore considered unavailable to cells. Albumin-bound testosterone represents approximately 50% of the total concentration. It is weakly bound and dissociates easily and rapidly. Testosterone bound to CBG represents 4% of the total T. Like albumin, it is weakly bound and dissociates rapidly. Free testosterone represents only about 2–3% of the total testosterone [21]. The term "bioavailable" testosterone refers to the sum of the CBG-bound, albumin-bound, and free components, and represents the testosterone fraction that is available to cells.

SHBG can vary considerably in men, greatly affecting total testosterone levels. Because SHBG-bound testosterone is not bioavailable, relying solely on total testosterone as an indicator of the adequacy of circulating androgens available for physiologic effect may lead to failure of diagnosing hypogonadism. Common medical conditions that increase SHBG include aging, hyperthyroidism, and hepatic cirrhosis. Those that decrease SHBG include obesity, diabetes mellitus, and glucocorticoid use. These are summarized in Table 2.1. Variability in SHBG levels makes the consideration of free and bioavailable testosterone important, in order to

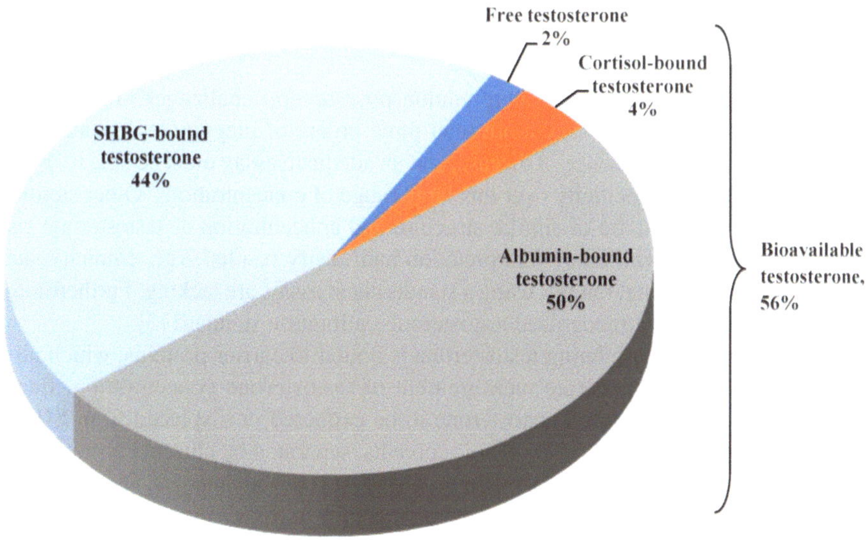

Fig. 2.1 Fractions of circulating total testosterone

Table 2.1 Factors affecting SHBG levels

Factors that increase SHBG	Factors that decrease SHBG
Aging	Obesity
Cirrhosis	Nephrotic syndrome
Hepatitis	Acromegaly
Hyperthyroidism	Hypothyroidism
Anticonvulsants	Glucocorticoids
Estrogens	Progestins
HIV disease	Androgenic steroids
	Diabetes mellitus

better appreciate the physiologically active circulating testosterone concentration. For example, the increase in SHBG with age means that older men may demonstrate a normal total testosterone level, even if they are hypogonadal, with low levels of free or bioavailable testosterone. Conversely, obesity decreases SHBG and total testosterone, even when the bioavailable fraction may be normal [4].

It should be noted that independent of SHBG levels, many of the features of the metabolic syndrome, such as hypertension, dyslipidemia, insulin resistance, and obesity, are commonly present in hypogonadal men [22, 23]. Hypogonadotropic hypogonadism occurs frequently in men with type II diabetes mellitus [24]. Expert opinion encourages the measurement of testosterone in men with metabolic syndrome and symptoms of testosterone deficiency, but the utility of androgen replacement therapy in these men continues to be an active area of discussion and research [3].

Laboratory Measurement of Testosterone

Testosterone assays and their interpretation pose several challenges. Testosterone concentrations in plasma vary more than three orders of magnitude depending on age, gender, and the presence of disease, and an adequate assay must be able to maintain sensitivity and specificity over this large range of concentrations. Other steroids in the circulation that are of similar structure and concentration as testosterone can lead to difficulties with assay interpretation and falsify results. Age, ethnicity, and gender-adjusted normal ranges, using a standardized assay, are lacking. Furthermore, there is no universally recognized testosterone-calibration standard [5].

The majority of circulating testosterone is bound to carrier proteins, which also has implications for accurate measurement of testosterone concentration. Early testosterone assays required testosterone to be extracted or displaced from SHBG and albumin by dissolution into organic solvents, separated by chromatography, and then measured (Fig. 2.2). This method offered several advantages, including the separation of interfering proteins and cross-reacting steroids to increase specificity, and the use of large serum aliquots to increase sensitivity. Unfortunately, this method was also time-intensive and expensive, and typically limited to research laboratories. With increased demand for economical and rapid alternatives, these early assays were replaced by radioimmunoassays (RIAs), which are presently the longest-used tests for the measurement of serum testosterone. Most of the current "normal ranges" are based on RIA results.

Over the last decade, the laboratory investigation of total testosterone measurement has evolved from RIA testing after serum extraction, to direct automated enzyme-linked immunoassay (ELI) testing available in most laboratories, to mass spectroscopy methods in reference laboratories. Each approach has its advantages and disadvantages. Correlation between these various methodologies can be poor, and national standardization is necessary in this area for portability and comparability of results obtained from the same patient.

Radioimmunoassay

RIAs are based on competitive binding of testosterone to a testosterone-specific antibody. The patient's serum is mixed with a set amount of radioactively labeled testosterone tracer and a fixed amount of antibody against testosterone. The amount of tracer displaced by the patient's testosterone is evaluated by measuring the radioactivity of the sample, and the patient's testosterone concentration is calculated. This method is illustrated in Fig. 2.3.

Traditional RIAs involve a testosterone extraction step, which increases their sensitivity and accuracy. These assays exhibit a variability of less than 10% in an experienced laboratory, with a lower limit of detection less than 10 pg [25]. Unfortunately, they require a technically experienced staff, are time-intensive, and

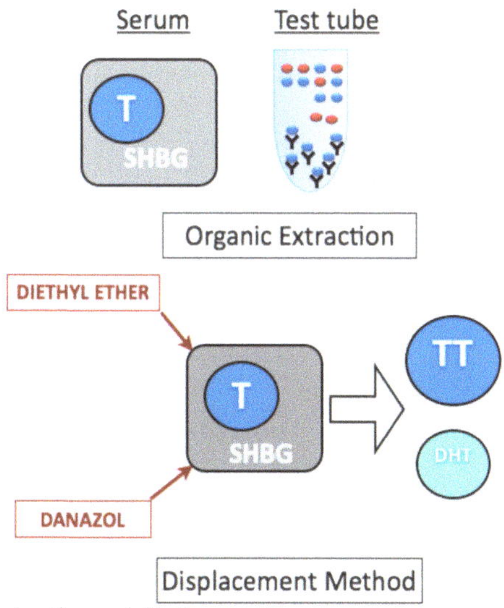

Fig. 2.2 Testosterone extraction and displacement

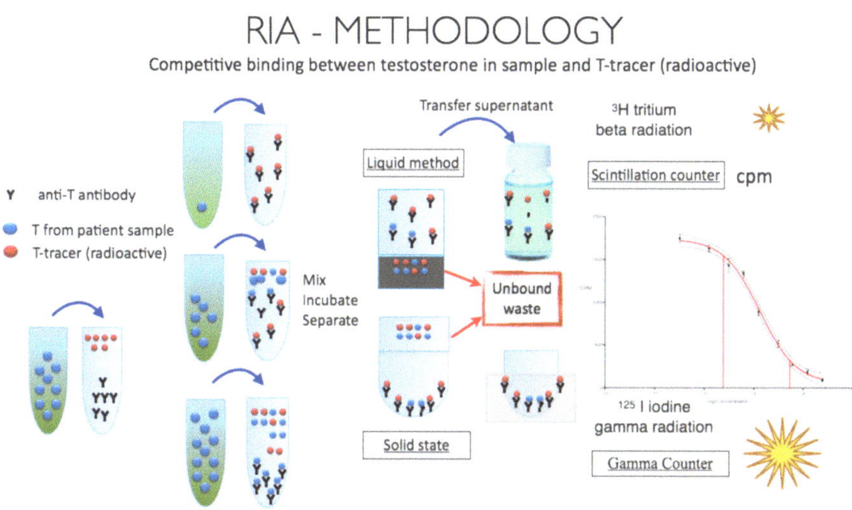

Fig. 2.3 Radioimmunoassay for measurement of serum testosterone

generate large amounts of radioactive waste. Commercially available RIA kits are an easier alternative. However, they use a set amount of tracer and antibody, with the result that their ability to measure very high or very low levels of testosterone is somewhat poor. The detection of low levels of testosterone can be improved by increasing the amount of antibody while reducing the amount of tracer [4]. This concept is illustrated in Fig. 2.4, which shows an immunoassay calibration curve. The testosterone concentration of a sample is accurately calculated over the dynamic range, represented by the linear portion of the curve. The sensitivity and specificity of the assay

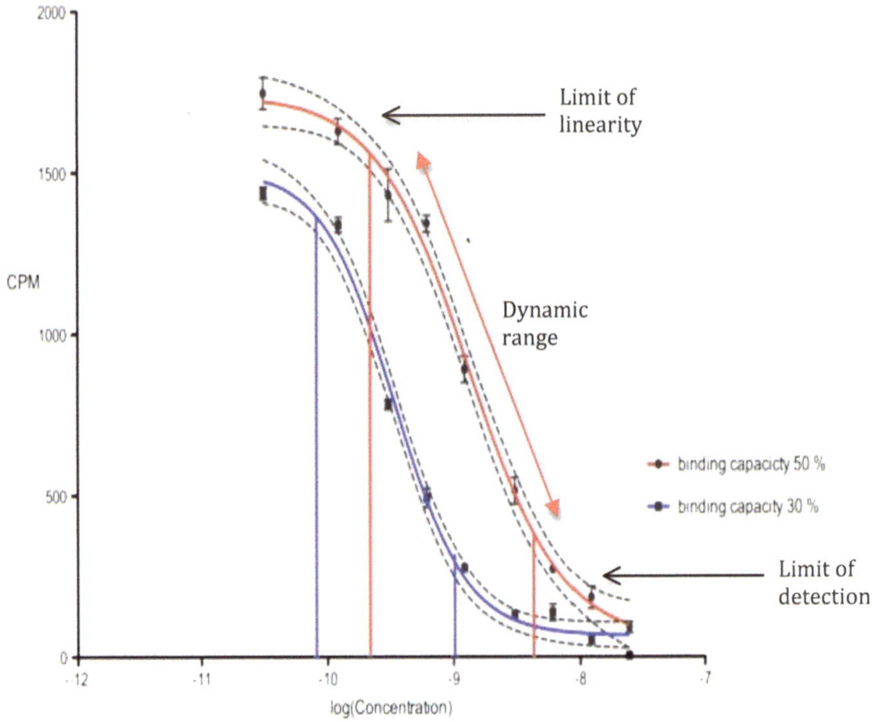

Fig. 2.4 Calibration curves for testosterone assays

is decreased at low testosterone concentrations, where the curve is no longer linear. Changes in the amount of antibody and/or tracer in the assay alter the binding capacity and shifts the curve, thereby affecting the dynamic range (Fig. 2.4).

The RIA requires a specific antibody with minimal cross-reactivity. Testosterone, like most steroids, is a poor antigen. Therefore, the design of a testosterone-specific antibody is, therefore, difficult. For commercially available kits, each manufacturer usually uses a different antibody. This may in turn cause a different binding affinity to testosterone, as well as different cross-reactivity. These differences can certainly contribute to the variable results seen with different commercial kits.

Enzyme-Linked Immunoassay

Like RIAs, EIAs are also based on the principle of competitive binding to an antibody. Instead of being radioactively labeled, the tracers used in EIA are fluorescent or chemiluminescent, which eliminates the possibility of radioactive waste. EIAs are easy to automate, and are widely used in non-reference hospital and commercial laboratories.

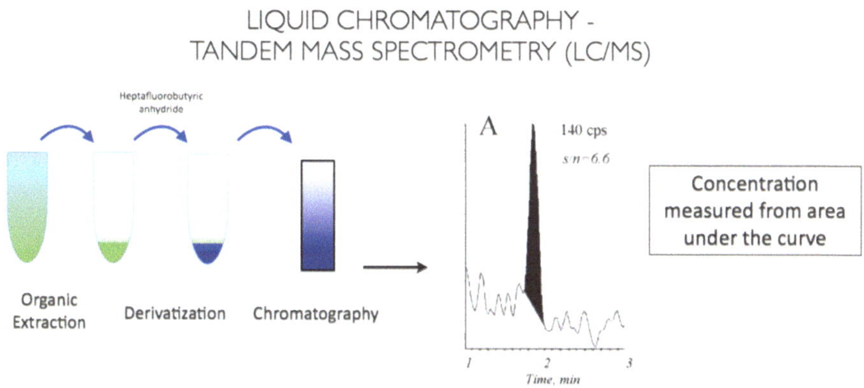

Fig. 2.5 Liquid chromatography-mass spectroscopy for measurement of serum testosterone

EIAs are subject to most of the same shortcomings of RIAs, namely, limited utility at very low or very high testosterone concentrations, and the need for a testosterone-specific antibody, which can vary by manufacturer. While these automated assays are sufficient for most clinical purposes in men, results can vary by as much as 33% among different assay methods. Even values in normal men can fall out of the manufacturer's reference range [14]. Additionally, EIAs do not routinely include a testosterone extraction step. Therefore, they do not necessarily measure true total testosterone.

Some manufacturers of commercial RIA and EIA use danazol for testosterone displacement as an alternative to testosterone extraction using organic solvents. In theory, this additional step is designed to improve the sensitivity and accuracy of the assays. However, the amount of danazol present in the kits is fixed and limited, and usually insufficient in displacing testosterone from samples with elevated amounts of SHBG.

Liquid Chromatography-Tandem Mass Spectroscopy

Direct methods of measuring testosterone have gained popularity over the last decade, due to the inconsistencies experienced with indirect immunoassays. Reference laboratories are increasingly using liquid chromatography-tandem mass spectroscopy (LC-MS), which combines automation and high-throughput with high precision, reproducibility, and wide linearity of assay. Steroid hormones are extracted from the serum sample by the use of organic solvents and separated by liquid chromatography, and testosterone levels are then determined by peak area integration of testosterone-containing fractions of the column eluate (Fig. 2.5). LC-MS has a lower limit of detection of 15 ng/dL, with an intra-assay coefficient of variation of <10% [14]. The specificity of this assay is enhanced by a distinct mass spectroscopy fingerprint for all known steroids, from which testosterone can be specifically selected. LC-MS is recognized as the most reliable method of measuring total testosterone in women, children, and hypogonadal men [26, 27], and has been accepted as the new gold standard assay for testosterone measurement.

Laboratory Measurement of Testosterone Fractions

Assays for various testosterone fractions have also been developed, and are summarized in Table 2.2. Some laboratories have directly measured bioavailable testosterone by selective ammonium sulfate precipitation of SHBG. Others have measured free testosterone by equilibrium dialysis, and advocated for this being the most accurate assessment of physiologically active circulating hormone fraction. The calculation of free testosterone, bioavailable testosterone, and the free androgen index (FAI = testosterone/SHBG) is described as an indirect means of assessing testosterone concentration.

According to the "free hormone hypothesis," only free testosterone is available to the end organs and biologically active at the tissue level [28]. However, recent studies have suggested that even SHBG-bound testosterone can be internalized into cells via endocytosis, questioning the validity of the free hormone hypothesis. Because of these conceptual uncertainties, while some authors continue to argue in favor of using free testosterone assays or calculations, expert opinion at this time favors the use of total testosterone levels for the diagnosis of androgen deficiency.

Measurement of Free Testosterone

Unfortunately, there is no cost-effective, simple, widely available, reliable, or accurate method of directly measuring free testosterone. The serum concentration of free testosterone is usually 50–100 times lower than the concentration of total

Table 2.2 Assays for the measurement of testosterone, free testosterone, and bioavailable testosterone

Total testosterone	Radioimmunoassay (RIA)
	Enzyme-linked immunoassay (EIA)
	Liquid chromatography-mass spectroscopy (LC-MS)
Free testosterone	Equilibrium dialysis
	Ultracentrifugation
	Radioimmunoassay (RIA)
	Calculated free testosterone
Bioavailable testosterone	Ammonium sulfate precipitation of SHBG
	Calculated bioavailable T

testosterone; therefore, lower limits of detection are important to consider for any assay measuring free T. The free hormone fraction also needs to be separated from bound testosterone in order to be measured.

Equilibrium dialysis is the gold standard assay for the measurement of free testosterone. The patient's serum is placed in a dialysis chamber with two compartments. One compartment contains 3 H-T (tracer-labeled testosterone) added to the serum, while the other contains a buffer solution. The compartments are separated by a semipermeable membrane with a low molecular weight cut-off. Dialysis is performed until testosterone concentrations within the two compartments reach an equilibrium value. Free and bound 3 H-T are then separated and measured. This value is multiplied by the total testosterone concentration. Although highly precise, equilibrium dialysis methods are expensive and complex, with potential errors due to temperature, sample dilution, and tracer impurities [14]. For example, a 2% contamination, with a tracer that does not bind to proteins can cause a doubling of the free testosterone concentration [5]. The value and accuracy of total testosterone has a bearing on the free testosterone concentration obtained with equilibrium dialysis.

One alternative to equilibrium dialysis for the direct measurement of free T is ultracentrifugation, a process where free T is separated from bound testosterone using ultrafast centrifuges, and then directly measured [29]. The direct measurement of free testosterone by RIA has also been described, but does not provide an accurate assessment when there are alterations in SHBG [30, 31]. The use of this technique is not recommended [32].

Calculation of Free Testosterone

Free testosterone can be calculated based on the concentration of SHBG, albumin, and total testosterone. Several different calculations for free T have been described in the literature [33]. These equations require the use of equilibrium dissociation constants for the binding of SHBG and albumin to testosterone, which are usually defined as 1×10^9 and 3×10^4 L/M, respectively. These numbers are not universally agreed upon, which can lead to discrepant results when the equations are compared [33]. The albumin concentration in serum is assumed to be 4.3 g/dL for the purposes of this calculation; changes in the albumin level within normal physiologic range have been shown to have little impact on the calculated free testosterone [30]. However, because the calculated value of free T depends on measured SHBG and total testosterone levels, measurement error and variable reference ranges for these two analytes directly impact the results. With the exception of certain metabolic conditions such as obesity, hyperthyroidism, and low cortisol, calculated free T correlates very well with free T obtained by equilibrium dialysis [30]. In contrast, studies comparing calculated values with those obtained by ultrafiltration do not always show high degree of correlation [34].

Free Androgen Index

The FAI represents the ratio of total testosterone and SHBG: testosterone/SHBG. The ratio is easy to calculate, and valid in serum samples from women. There is a reasonable correlation between FAI and FT, but it is not consistently maintained at low testosterone levels [35]. FAI is less useful in men because the majority of SHBG in men is bound to testosterone. Like the calculated free testosterone, the FAI is also dependent on accurate values for testosterone and SHBG.

Measurement of Bioavailable Testosterone

Bioavailable testosterone can be determined by ammonium sulfate precipitation of SHBG. 3 H-T (tracer-labeled testosterone) is mixed with the serum sample and SHBG-bound 3 H-T is then selectively precipitated by adding 50% ammonium sulfate. The 3 H-T that remains in the supernatant (not precipitated) is multiplied by the total testosterone concentration in order to determine the concentration of bioavailable testosterone. The assay is labor-intensive and does not lend itself easily to automation. It involves a series of steps, requires special reagents, and has not been standardized between labs. Despite these shortcomings, bioavailable T has been reported to correlate with free T levels obtained by equilibrium dialysis relatively well [30, 36].

Calculation of Bioavailable Testosterone

There is a highly positive correlation between the levels of calculated free T and bioavailable T in plasma. The two fractions are also functionally correlated; in cross-sectional studies of healthy older men, both free T and bioavailable T predict muscle mass strength and bone density [14].

Bioavailable T can be calculated from SHBG and total testosterone. It has been recommended by one expert panel as being superior, both in ease of execution and accuracy of results, to the ammonium-sulfate precipitation method for measurement of bioavailable T [37]. Several different equations for the calculation of bioavailable T have been proposed, but comparison of these algorithms suggests that the Sodergard or Vermeulen equations most accurately reflect measured bioavailable T levels [20].

Calculated values of bioavailable T are dependent on the determinations of SHBG and total testosterone, and therefore, subject to the same bias as calculated values of free T. The two measures are directly linked to each other [30]. An online calculator, available on the website for the International Society for the Study of the Aging Male (http://www.issam.ch/freetesto.htm), can be used for the calculation of both free and bioavailable testosterone.

Salivary Testosterone

There is a positive correlation between total testosterone in serum and saliva [38]. Salivary samples are easily collected and are useful research tools for settings in which blood sampling is impractical or impossible. Salivary testosterone measurements, however, are not standardized. Adult male reference ranges are not available in most hospital or reference laboratories [32]. This method of testing is not currently recommended for clinical use.

Internal and External Validity of Testosterone Assays

An accurate diagnosis of androgen deficiency requires hormone assays that have good internal and external validity. Internal validity implies that the assay has a high level of precision with low bias. External validity refers to calibration of the assay against the accepted gold standard with appropriate reference ranges [39]. Historically, testosterone assays, like other reproductive hormone assays, were established in research laboratories that maintained internal and external quality controls for the assays. The increasing use of these assays in routine clinical practice has increased the demand for them, leading to a switch from research laboratories to large-throughput laboratories that employ commercial, multiplex, automated platform assays.

The methodological simplifications required to make the transition to automation and high-throughput have led to a loss in sensitivity and specificity for testosterone assays. Widely differing reference ranges have been reported, both among the various automated assays, as well as compared to research laboratory methods. Reference ranges for many automated assays span low-end clinical detection limits of 170–200 ng/dL to upper-end limits of 700–800 ng/dL. This is significantly lower than the 300–1,000 ng/dL reference range published over the past 30 years based on RIAs with or without chromatography [32].

There is no uniform agreement between commonly used commercial assay platforms and traditional testosterone assays based extraction and chromatography. Wang et al. compared serum testosterone measurements from eugonadal and hypogonadal adult men using LC-MS vs. manual RIAs and four commonly used commercially available RIA kits [32]. Using LC-MS as the gold standard, they found that while some of the manual and automated assays could be used to assess testosterone in eugonadal men, the majority were unacceptable for measurement of low testosterone levels (i.e., hypogonadal men), due to lack of precision and accuracy. Similar findings have been reported by Taieb et al., who found that immunoassay results varied by up to fivefold at testosterone concentrations below 230 ng/dL, and that immunoassays generally overestimated testosterone concentrations [40]. Validation studies of most testosterone assays are performed using a standard sample with a testosterone concentration of 400 ng/dL. The sensitivity and specificity of most assays is not measured at low testosterone levels, where the linearity of the assay is decreased (Fig. 2.4). At low testosterone levels, as seen in women, children, and early male puberty, blood testosterone levels are comparable to those seen in castrate men. The low reliability and poor sensitivity of commercial testosterone assays in these settings has led to measurements being considered equivalent to random number generation [41].

Wang et al. recommended that laboratories employing automated assays should establish their own reference ranges, based on normal healthy men of different ages, rather than using manufacturer-supplied ranges [32]. Reference ranges for commercial testosterone assays are often based on small, convenient population samples, not necessarily controlled for medical comorbidities or other factors that may affect sex hormone levels [42]. Not surprisingly, reference ranges vary significantly between different commercial assays. Historically, the range of testosterone in healthy young men using assays that utilized extraction and chromatography has approximated 225–1,000 ng/dL. In contrast, some commercial laboratories have reported the lower limit of the normal range to be as low as 84 ng/dL and as high as 1,727 ng/dL, values that defy credibility and validity [42]. An acceptable reference range for testosterone assays should, instead, be based on a population of healthy men with verified normal sexual and reproductive function. However, to our knowledge, no large studies of a random sample of men from a healthy, population-based cohort exist.

Laboratories should also undertake external quality control programs, such as that provided by the College of American Pathologists, allowing laboratories to compare results with others using the same kit or reagents. In one such quality control study, the average value for serum testosterone was 297 ng/dL, with a range of 160–508 ng/dL, and coefficients of variation ranging from 5.1 to 22.7% [32]. Results clearly spanned the hypogonadal to eugonadal range. This magnitude of variability is unacceptable and of no value in clinical medicine. As previously mentioned, pre-analytical processing also influences the reliability of testosterone assays. Recognition of technical and biological pre-analytical issues can limit the potential for introduction of errors, and decrease inter-assay variability.

The laboratory diagnosis of testosterone deficiency may, therefore, be confounded by limitations of both internal and external validation. An understanding of the assays employed by one's specific laboratory, as well as an appreciation for population-specific reference ranges, is key for accurate clinical diagnosis.

Summary

Reproducible and cost-effective testosterone assays that reflect accurate serum concentrations are important in order to correctly diagnose hypogonadism. Over the past 30 years, testosterone assays have progressively become more economical, rapid, and automated. However, scientific data suggests that most testosterone measurements in typical clinical laboratories may have a margin of error less than or equal to 30% from the "true" serum testosterone concentrations measured, using gold standard methods such as liquid chromatography-mass spectroscopy [4].

Needless to say, the accuracy of measurement is especially important in hypogonadal men, where a 30% variation in testosterone concentrations may effectively miss a diagnosis of testosterone deficiency. The poor reliability of testosterone measurement is secondary to a combination of factors, including technical limitations of currently available assays, intra-individual variation, and the wide range of testosterone levels in human samples that assays are required to measure.

There is no universally accepted threshold of testosterone concentration that distinguishes eugonadal from hypogonadal men, but a testosterone concentration below 300 ng/dL is suggestive of testosterone deficiency in the appropriate clinical setting. The laboratory evaluation of testosterone deficiency should begin with the measurement of total testosterone, and may include free or bioavailable testosterone if the total testosterone is close to the lower limit of the normal range or discordant with the clinical presentation, or when altered levels of SHBG are suspected. LC-MS is considered the current gold standard testosterone assay. Determination of free testosterone should be done with equilibrium dialysis, or calculated after accurate determination of SHBG. Assessment of bioavailable testosterone is an alternative to the assessment of free testosterone, and can either be calculated or determined by ammonium-sulfate precipitation of SHBG. Both equilibrium dialysis and ammonium-sulfate precipitation are labor-intensive assays requiring skilled personnel, and are best performed in reference laboratories only.

Currently, efforts to create standardization of testosterone assays are underway. Accurate reference ranges for testosterone measurement by liquid chromatography-tandem mass spectroscopy are being developed. These norms need to be established in normal healthy patients of varying ages, with normal sexual and reproductive function. Calibration of the methodology and population-based reference ranges for free testosterone by equilibrium dialysis are also needed. Increased accuracy in the measurement of testosterone and SHBG will lead to improvement in calculations of free- and bioavailable testosterone.

In summary, it is in everyone's best interest to advance the technological and clinical aspects of biochemical diagnosis of hypogonadism, in order to provide reliable, cost-effective, and portable tools to aid clinicians in the diagnosis and treatment of this common condition.

Acknowledgements This work was partially supported by Robert Dow Foundation and Mr. and Mrs. Howard and Irena Laks Foundation unrestricted educational grants. We thank Ms. Kristin Saunders for editorial support.

References

1. Vesper HW, Botelho JC, Shacklady C, et al. CDC project on standardizing steroid hormone measurements. Steroids. 2008;73:1286–92.
2. Bhasin S, Cunningham G, Hayes FJ, et al. Testosterone therapy in men with androgen deficiency syndromes: an Endocrine Society clinical practice guideline. J Clin Endocrinol Metab. 2010; 95:2536–99.

3. Wang C, Nieschlag E, Swerdloff R, et al. Investigation, treatment, and monitoring of late-onset hypogonadism in males: ISA, ISSAM, EAU, EAA, and ASA recommendations. J Androl. 2009;30:1–9.
4. Hellstrom JG, Paduch D, Donatuci CF. Importance of hypogonadism and testosterone replacement in current urologic practice: a review. Int Urol Nephrol. 2012;44:61–70.
5. Rosner W, Auchis RJ, Azziz R, et al. Position statement: utility, limitations, and pitfalls in measuring testosterone: an Endocrine Society position statement. J Clin Endocrinol Metab. 2007;92:405–13.
6. Araujo AB, Esche GR, Kupelian V, et al. Prevalence of symptomatic androgen deficiency in men. J Clin Endocrinol Metab. 2007;92:4241–7.
7. Harman SM, Metter EJ, Tobin JD, et al. Longitudinal effects of aging on serum total and free testosterone levels in healthy men. J Clin Endocrinol Metab. 2001;86:724–31.
8. Wu FCW, Tajar A, Beynon J, et al. Identification of late-onset hypogonadism in middle-aged and elderly men. N Eng J Med. 2010;363:123–35.
9. Feldman HA, Longcope C, Derby CA, et al. Age trends in the level of serum testosterone and other hormones in middle-aged men: longitudinal results from the Massachusetts male aging study. Clin Endocrinol Metab. 2002;87:589–98.
10. Gray A, Feldman HA, McKinlay JB, Longcope C. Age, disease and changing sex hormone levels in middle-aged men: results of the Massachusetts male aging study. J Clin Endocrinol Metab. 1991;73:1016–25.
11. Cunningham GR, Toma SM. Why is androgen replacement in males controversial? J Clin Endocrinol Metab. 2011;96:38–52.
12. Brambilla DJ, Matsumoto AM, Araujo AB, McKinlay JB. The effect of diurnal variation on clinical measurement of serum testosterone and other sex hormone levels in men. J Clin Endocrinol Metab. 2009;94:907–13.
13. Plymate SR, Tenover JS, Bremner WJ. Circadian variation in testosterone, sex-hormone binding globulin, and calculated non-sex hormone binding globulin bound testosterone in healthy young and elderly men. J Androl. 1989;10:366–71.
14. Winters SJ. Laboratory assessment of testicular function. In: Endotext. http://www.endotext.org/male/male4/male4.html. Accessed 24 Apr 2012.
15. Diver MJ, Imtiaz KE, Ahmad AM, et al. Diurnal rhythms of serum total, free, and bioavailable testosterone and of SHBG in middle-aged men, compared with those in younger men. Clin Endocrinol. 2003;58:710–7.
16. Bremner WJ, Vitiello MV, Prinz PN. Loss of circadian rythmicity in blood testosterone levels with aging in normal men. J Clin Endocrinol Metab. 1983;56:1278–81.
17. Brambilla DJ, O'Donnell AB, Matsumoto AM, McKinlay JB. Intra-individual variation in levels of serum testosterone and other reproductive and adrenal hormones in men. Clin Endocrinol. 2007;67:853–62.
18. Raff H, Sluss PM. Pre-analytical issues for testosterone and estradiol assays. Steroids. 2008;73:1297–304.
19. Bolyakov A, Raymond S, Vaucher L, et al. Changes in serum in reproductive hormones and neurotransmitters during arousal, ejaculation, and orgasm in normal volunteers. J Urol. 2010;183:468–9.
20. Dunn JF, Nisula BC, Rodbard D. Transport of steroid hormones: binding of 21 endogenous steroids to both testosterone-binding globulin and corticosteroid-binding globulin in human plasma. J Clin Endocrinol Metab. 1981;53(1):58–68.
21. De Ronde W, van der Schouw YT, Pols HAP, et al. Calculation of bioavailable and free testosterone in men: a comparison of 5 published algorithms. Clin Chem. 2006;52:1777–84.
22. Kupelian V, Page ST, Araujo AB, et al. Low sex-hormone binding globulin, total testosterone, and symptomatic androgen deficiency are associated with development of the metabolic syndrome in nonobese men. J Clin Endocrinol Metab. 2006;91:843–50.
23. Kalyani RR, Dobs AS. Androgen deficiency, diabetes, and the metabolic syndrome in men. Curr Opin Endocrinol Diabetes Obes. 2007;14:226–34.
24. Dhindsa S, Prabhakar S, Sethi M, et al. Frequent occurrence of hypogonadotropic hypogonadism in type 2 diabetes. J Clin Endocrinol Metab. 2004;89:5462–8.

25. Nieschlag E, Loriaux DL. Radioimmunoassay for plasma testosterone. Z Klin Chem Klin Biochem. 1972;10:164–8.
26. Albrecht L, Styne D. Laboratory testing of gonadal steroids in children. Pediatr Endocrinol Rev. 2007;5 Suppl 1:599–607.
27. Bui HN, Struys EA, Martens F, et al. Serum testosterone levels measured by isotope dilution-liquid chromatography-tandem mass spectrometry in postmenopausal women versus those in women who underwent bilateral oophorectomy. Ann Clin Biochem. 2010;47(Pt 3):248–52.
28. Herzog AG, Levesque LA. Testosterone, free testosterone, nonsex hormone binding globulin, and free androgen index: which testosterone measurement is most relevant to reproduction and sexual function in men with epilepsy? Arch Neurol. 1992;49:133–5.
29. Chen Y, Yazdanpanah M, Wang XY, et al. Direct measurement of serum free testosterone by ultrafiltration followed by liquid chromatography tandem mass spectrometry. Clin Biochem. 2020;43:490–6.
30. Vermeulen A, Verdonck L, Kaufman JM. A critical evaluation of simple methods for the estimation of free testosterone in serum. J Clin Endocrinol Metab. 1999;84:3666–72.
31. Rosner W. Errors in the measurement of plasma free testosterone. J Clin Endocrinol Metab. 1997;82:2014–5.
32. Wang C, Caitlin DH, Demers LM, et al. Measurement of total serum testosterone in adult men: comparison of current laboratory methods versus liquid chromatography-tandem mass spectroscopy. J Clin Endocrinol Metab. 2004;89:534–43.
33. Sartorius G, Ly LP, Sikaris K, et al. Predictive accuracy and sources of variability in calculated free testosterone estimates. Ann Clin Biochem. 2009;46:137–43.
34. Van Uytfanghe K, Stockl D, Kaufman JM, et al. Evaluation of a candidate reference measurement procedure for serum free testosterone based on ultrafiltration and isotope-dilution gas chromatography–mass spectrometry. Clin Chem. 2004;50:2101–10.
35. Kapoor P, Luttrell BM, Williams D. The free androgen index is not valid for adult males. J Steroid Biochem. 1993;45:325–6.
36. Morley JE, Patrick P, Perry HM. Evaluation of assays available to measure free testosterone. Metabolism. 2002;51:554–9.
37. Collier CP, Clark AF, Bain J, et al. Functional testosterone: biochemical assessment of hypogonadism in men—report from a multidisciplinary workshop hosted by the Ontario Society of Clinical Chemists. Aging Male. 2007;10:211–6.
38. Wang C, Plymate S, Nieschlag E, Paulsen CA. Salivary testosterone in men: further evidence of a direct correlation with free serum testosterone. J Clin Endocrinol Metab. 1981;53:1021–4.
39. Sikaris K, McLachlan RI, Kazlauskas R, et al. Reproductive hormone reference intervals for healthy fertile young men: evaluation of automated platform assays. J Clin Endocrinol Metab. 2005;90:5928–36.
40. Taieb J, Mathian B, Millot F, et al. Testosterone measured by 10 immunoassays and by isotope-dilution gas chromatography–mass spectrometry in sera from 116 men, women, and children. Clin Chem. 2003;49:1381–95.
41. Herold DA, Fitzgerald RL. Immunoassays for testosterone in women: better than a guess? Clin Chem. 2003;49:1250–1.
42. Bhasin S, Zhang A, Coviello A, et al. The impact of assay quality and reference ranges on clinical decision making in the diagnosis of androgen disorders. Steroids. 2008;73:1311–7.

Chapter 3
The Clinical Diagnosis of Androgen Deficiency

Gregory C. Mitchell, Ege Can Serefoglu, and Wayne J.G. Hellstrom

Introduction

The clinical diagnosis of androgen deficiency can often be a challenge for physicians. Its presentation can vary and be as diverse as the effects that testosterone has on different body systems (see Table 3.1). The severity of the signs and symptoms between body organs is often discordant, with the magnitude of signs and symptoms often not matching measured testosterone levels. Although the true diagnosis of androgen deficiency requires confirmatory laboratory testing of serum testosterone levels [1], clinicians must often use their best clinical judgment about when to order laboratory tests, since there is insufficient evidence to justify screening for androgen deficiency in the general population [2]. The currently accepted solution is to consider subjective symptoms, along with observed signs, and allow those metrics to guide the judicious use of diagnostic testing [2]. Patients need to be questioned about recognized causes of testicular failure; such as trauma, testicular cancer, chemotherapy, radiation exposure, and orchitis. In addition to a complete history and physical examination, clinicians may use one of the validated questionnaires to help identify men in need of further endocrinologic evaluation. The Aging Males Symptoms (AMS) scale [3], the androgen deficiency in aging males (ADAM) survey [4], the Androtest interview [5], and the International Index of Erectile Function (IIEF) [6] are used as helpful adjuncts to the history and physical examination in the clinical diagnosis of androgen deficiency. The use of these instruments is reviewed in greater detail in Chap. 4.

G.C. Mitchell, MS • E.C. Serefoglu, MD • W.J.G. Hellstrom, MD, FACS (✉)
Department of Urology, Tulane University Medical Center, 1430 Tulane Avenue
SL-42, New Orleans, LA 70112, USA
e-mail: whellst@tulane.edu

W.J.G. Hellstrom (ed.), *Androgen Deficiency and Testosterone Replacement: Current Controversies and Strategies*, Current Clinical Urology,
DOI 10.1007/978-1-62703-179-0_3,

Table 3.1 Summary of symptoms and signs of androgen deficiency

Group A: Symptoms and signs suggestive of androgen deficiency in men
• Incomplete sexual development, aspermia, eunuchoid body habitus
• Reduced sexual desire (libido) and activity
• Decreased spontaneous erections
• Breast discomfort, gynecomastia
• Loss of body (auxiliary and pubic) hair, reduced shaving
• Very small or shrinking testis (especially <5 cm^3)
• Inability to father children, reduced or absent sperm in ejaculate
• Height loss, soft-trauma fracture, low bone mineral density, osteomalacia
• Reduced muscle mass and strength
• Hot flashes, sweats
Group B: Symptoms and signs associated with androgen deficiency that are less specific than those listed in group A
• Decreased energy, motivation, initiative, aggressiveness, self confidence
• Feeling sad or down, depressed mood, dysthymia
• Impaired concentration and memory
• Prolonged sleep disturbance, increased sleepiness
• Mild anemia (normochromic, normocytic, into the female range)
• Increased body fat, body mass index
• Diminished physical or work performance

Based on data from Bhasin et al. [2]

Presentation by Age

The clinical picture of male hypogonadism varies markedly between patients who have yet to undergo puberty and those subjects that have already reached sexual maturity. It behooves the clinician to be astute regarding certain signs and symptoms that occur at each stage of life. In the pubescent male, the most noticeable signs of reduced testosterone are a lack of age-appropriate virilization and a paucity of secondary sexual characteristics, such as deepening of the voice and temporal hair recession [7]. These patients can display sustained height increase (albeit without a pubertal growth spurt) because of a failure of the epiphyses to close, and exhibit eunuchoid body proportions, where arm span is greater than overall height by >2 cm [8].

Hypogonadism in the pubescent male can indicate of a genetic disorder. In these cases, the classic signs and symptoms of these rare disorders can alert the clinician to the presence of a coincident hypogonadal state. In Kallmann syndrome, low serum levels of testosterone result from reduced or absent gonadotropin releasing hormone (GnRH), and juvenile boys may present with anosmia, delayed sexual development, and midline facial/cranial defects [9]. Although Klinefelter syndrome is the most common genetic cause of hypogonadism, patients typically exhibit a clinically normal phenotype until puberty. At which point, serum LH and FSH levels dramatically increase, inhibin B decreases, and testosterone levels fall, leading to hypogonadal manifestations [10]. Pre-pubertal Klinefelter boys have preserved normal seminiferous tubule architecture with characteristically reduced germ cell counts. In contrast, in Klinefelter patients who have undergone puberty, their seminiferous tubules undergo extensive hyalinization and fibrosis with testicular atrophy [10]. Bilateral congenital anorchia is another potential genetically linked etiology of testosterone deficiency that is easily identified by congenital absence of the testes [11].

Patients who have reached sexual maturity exhibit a different constellation of symptoms, mainly centering on a loss of secondary sexual characteristics. Loss of body hair with a reduced frequency of shaving, decreased muscle mass, physical strength, and atrophic testicles are typical examples [2]. According to the Endocrine Society, other specific symptoms of testosterone deficiency include reduced sexual desire and frequency of sexual activity, decreased spontaneous and nocturnal erections, hot flashes, and breast tenderness or gynecomastia [2].

Often, the symptoms of testosterone deficiency are difficult to distinguish from the physiologic changes that we witness with normal aging. Therefore, testosterone deficiency in the aging male is often not recognized [12]. There are also differences in the presentation of testosterone deficiency in middle-aged men from the geriatric population. Many signs and symptoms of testosterone deficiency are expressed heterogeneously with respect to age in the post-pubertal population [13]. In a series of 1,647 males who were assayed for signs and symptoms of testosterone deficiency, the youngest quartile showed an inverse correlation between testosterone and hypoactive sexual desire, while the oldest quartile demonstrated inverse correlations between testosterone and both severe erectile dysfunction and somatized anxiety [13].

Signs and Symptoms of Androgen Deficiency by System

Sexual Dysfunction

The hallmark of androgen deficiency to the general population is diminished sexual function (see Fig. 3.1). Reduced libido is often observed in men with undersized or soft testicles and erectile dysfunction (ED) [7, 14]. Other sexual dysfunctions associated with androgen deficiency include decreased erectile rigidity, difficulty achieving orgasm, diminished ejaculatory volume, reduced genital sensation, and diminished intensity of orgasm [14].

Hypogonadal men with low libido and ED have been observed to have long-term improvement in their sexual desire with exogenous testosterone supplementation. However, their resolution of ED is short lived, when measured using validated questionnaires [15, 16]. Nevertheless, it is important to recognize a correlation between ED and testosterone deficiency, and that a new onset of ED can be a harbinger of future vascular morbidity and mortality [17].

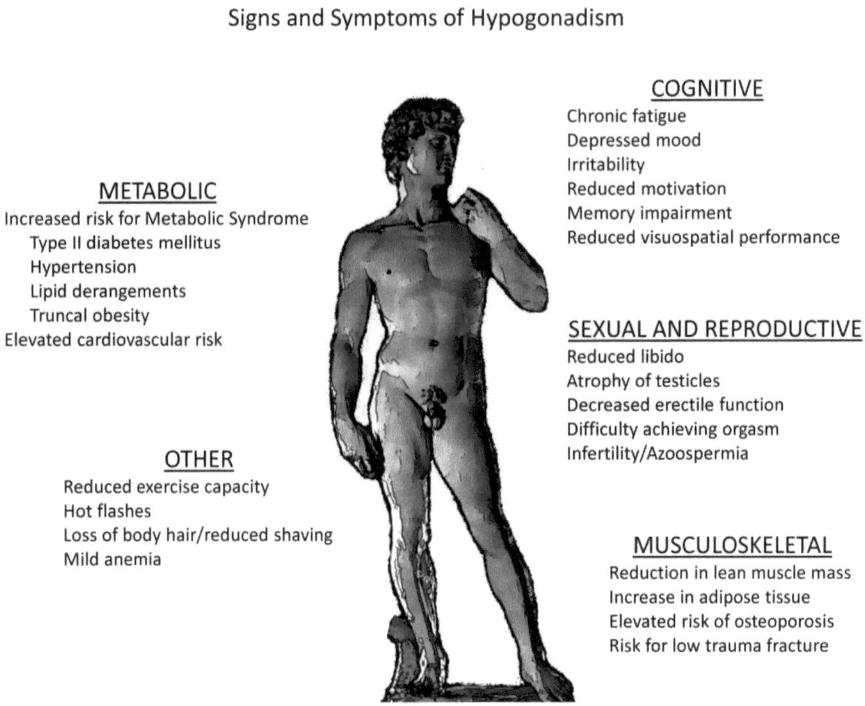

Fig. 3.1 Summary of signs of testosterone deficiency by system

Cognition and Mood Changes

Chronic fatigue and lack of energy are the most common mood abnormalities associated with testosterone deficiency [18]. Depression, irritability, and reduced motivation are common entities witnessed in the hypogonadal state [2, 19]. Sleep disturbance and decreased work capacity are nonspecific symptoms that are often associated with reduced testosterone [2]. Other cognitive effects seen in men with subnormal testosterone levels include memory impairment and worsening visuospatial performance [20].

Metabolic Syndrome

The metabolic syndrome (MetS) is a group of coincident disease processes that increases an individuals' risk of cardiovascular events although multiple definition and content have been proposed. According to the International Diabetes Federation, the major entities that comprise the MetS are hypertension, insulin resistance, abnormal body fat distribution, and dyslipidemia [21]. Studies have clearly noted an association between testosterone deficiency and the MetS [22, 23], an association that persists in non-obese males [24]. Cardiovascular complications associated with MetS and low testosterone include large vessel atherosclerosis [25] and angina pectoris [26]. Since testosterone therapy has been shown to be more effective than exercise and diet alone in treating MetS among patients with subnormal testosterone [27], it is vital to determine if the MetS patient is a candidate for testosterone replacement therapy [18].

Body Composition

There is an annual 1.5–2% reduction of testosterone production during the normal aging process in adult men [28]. Likewise, there is decrease in muscle mass and an associated increase in adipose tissue, which alters overall body shape and composition [28]. Studies of men with hypogonadism reveal decreases in lean body mass and increases in overall body fat mass paralleling falling androgen levels [29] (Fig. 3.1).

In addition to its effects on overall body composition, testosterone has an effect on regional body distributions of various tissues. Analysis of a cohort of men, aged 50–89, from the California community of Rancho-Bernardo documented that total testosterone levels were inversely correlated with central adiposity [30]. Testosterone replacement therapy in hypogonadal men increased lean body mass and decreased the total amount of adipose tissue while simultaneously altering its distribution [31].

Skeletal Changes

Testosterone deficiency contributes to reduced bone mineral density and the development of osteoporosis by interrupting the normal inhibition of osteoclastic bone resorption and decreasing the estrogen-dependent stimulation of osteoblasts, by way of the inability of testosterone to be aromatized into estrogen [32]. Low serum testosterone levels correlate with frailty because of decreased bone mineral density in elderly males [33].

Conclusion

The most important objective during the clinical assessment of the man with potential testosterone deficiency is to guide the clinician when to order confirmatory diagnostic testing. The Endocrine Society guidelines recommend measuring morning serum testosterone levels in men exhibiting any of the symptoms listed in group A of Table 3.1 [2]. The proper use of a detailed history and physical exam is the ideal approach to identify the man who is suffering from hypogonadism. Nevertheless, it is important for the clinician to detect the appropriate nuances and develop standards to decide when to screen for testosterone deficiency.

References

1. Wu FC, Tajar A, Beynon JM, et al. Identification of late-onset hypogonadism in middle-aged and elderly men. N Engl J Med. 2010;363(2):123–35.
2. Bhasin S, Cunningham GR, Hayes FJ, et al. Testosterone therapy in adult men with androgen deficiency syndromes: an Endocrine Society clinical practice guideline. J Clin Endocrinol Metab. 2006;91(6):1995–2010.
3. Heinemann LA, Saad F, Zimmermann T, et al. The Aging Males' Symptoms (AMS) scale: update and compilation of international versions. Health Qual Life Outcomes. 2003;1:15.
4. Morley JE, Charlton E, Patrick P, et al. Validation of a screening questionnaire for androgen deficiency in aging males. Metabolism. 2000;49(9):1239–42.
5. Corona G, Mannucci E, Petrone L, et al. ANDROTEST: a structured interview for the screening of hypogonadism in patients with sexual dysfunction. J Sex Med. 2006;3(4):706–15.
6. Rosen RC, Riley A, Wagner G, Osterloh IH, Kirkpatrick J, Mishra A. The international index of erectile function (IIEF): a multidimensional scale for assessment of erectile dysfunction. Urology. 1997;49(6):822–30.
7. Bhasin S, Basaria S. Diagnosis and treatment of hypogonadism in men. Best Pract Res Clin Endocrinol Metab. 2011;25(2):251–70.
8. Arver S, Lehtihet M. Current guidelines for the diagnosis of testosterone deficiency. Front Horm Res. 2009;37:5–20.
9. Pallais JC, Au M, Pitteloud N, Seminara S, Crowley WF. Kallmann syndrome. In: Pagon RA, Bird TD, Dolan CR, Stephens K, Adam MP, editors. GeneReviews™ [Internet]. Seattle (WA): University of Washington, Seattle; 1993–2007.
10. Wikstrom AM, Dunkel L. Klinefelter syndrome. Best Pract Res Clin Endocrinol Metab. 2011;25(2):239–50.
11. Brauner R, Neve M, Allali S, et al. Clinical, biological and genetic analysis of anorchia in 26 boys. PLoS One. 2011;6(8):e23292.
12. Pinsky MR, Hellstrom WJ. Hypogonadism, ADAM, and hormone replacement. Ther Adv Urol. 2010;2(3):99–104.
13. Corona G, Mannucci E, Ricca V, et al. The age-related decline of testosterone is associated with different specific symptoms and signs in patients with sexual dysfunction. Int J Androl. 2009;32(6):720–8.
14. Hellstrom WJ, Paduch D, Donatucci CF. Importance of hypogonadism and testosterone replacement therapy in current urologic practice: a review. Int Urol Nephrol. 2012; 44(1):61–70.
15. Mulhall JP, Valenzuela R, Aviv N, Parker M. Effect of testosterone supplementation on sexual function in hypogonadal men with erectile dysfunction. Urology. 2004;63(2):348–52.
16. Shabsigh R, Kaufman JM, Steidle C, Padma-Nathan H. Randomized study of testosterone gel as adjunctive therapy to sildenafil in hypogonadal men with erectile dysfunction who do not respond to sildenafil alone. J Urol. 2004;172(2):658–63.
17. Costa C, Virag R. The endothelial-erectile dysfunction connection: an essential update. J Sex Med. 2009;6(9):2390–404.
18. Traish AM, Miner MM, Morgentaler A, Zitzmann M. Testosterone deficiency. Am J Med. 2011;124(7):578–87.
19. Barrett-Connor E, Von Muhlen DG, Kritz-Silverstein D. Bioavailable testosterone and depressed mood in older men: the Rancho Bernardo study. J Clin Endocrinol Metab. 1999; 84(2):573–7.

20. Moffat SD, Zonderman AB, Metter EJ, Blackman MR, Harman SM, Resnick SM. Longitudinal assessment of serum free testosterone concentration predicts memory performance and cognitive status in elderly men. J Clin Endocrinol Metab. 2002;87(11):5001–7.
21. Alberti KG, Zimmet P, Shaw J. Metabolic syndrome—a new world-wide definition. A Consensus Statement from the International Diabetes Federation. Diabet Med. 2006;23(5):469–80.
22. Laaksonen DE, Niskanen L, Punnonen K, et al. Sex hormones, inflammation and the metabolic syndrome: a population-based study. Eur J Endocrinol. 2003;149(6):601–8.
23. Somani B, Khan S, Donat R. Screening for metabolic syndrome and testosterone deficiency in patients with erectile dysfunction: results from the first UK prospective study. BJU Int. 2010;106(5):688–90.
24. Kupelian V, Page ST, Araujo AB, Travison TG, Bremner WJ, McKinlay JB. Low sex hormone-binding globulin, total testosterone, and symptomatic androgen deficiency are associated with development of the metabolic syndrome in nonobese men. J Clin Endocrinol Metab. 2006;91(3):843–50.
25. Hak AE, Witteman JC, de Jong FH, Geerlings MI, Hofman A, Pols HA. Low levels of endogenous androgens increase the risk of atherosclerosis in elderly men: the Rotterdam study. J Clin Endocrinol Metab. 2002;87(8):3632–9.
26. Malkin CJ, Pugh PJ, Morris PD, et al. Testosterone replacement in hypogonadal men with angina improves ischaemic threshold and quality of life. Heart. 2004;90(8):871–6.
27. Heufelder AE, Saad F, Bunck MC, Gooren L. Fifty-two-week treatment with diet and exercise plus transdermal testosterone reverses the metabolic syndrome and improves glycemic control in men with newly diagnosed type 2 diabetes and subnormal plasma testosterone. J Androl. 2009;30(6):726–33.
28. Vermeulen A. Androgen secretion after age 50 in both sexes. Horm Res. 1983;18(1–3):37–42.
29. Bhasin S. Regulation of body composition by androgens. J Endocrinol Invest. 2003; 26(9):814–22.
30. Khaw KT, Barrett-Connor E. Lower endogenous androgens predict central adiposity in men. Ann Epidemiol. 1992;2(5):675–82.
31. Wang C, Swerdloff RS, Iranmanesh A, et al. Transdermal testosterone gel improves sexual function, mood, muscle strength, and body composition parameters in hypogonadal men. J Clin Endocrinol Metab. 2000;85(8):2839–53.
32. Michael H, Harkonen PL, Vaananen HK, Hentunen TA. Estrogen and testosterone use different cellular pathways to inhibit osteoclastogenesis and bone resorption. J Bone Miner Res. 2005;20(12):2224–32.
33. Bassil N. Late-onset hypogonadism. Med Clin North Am. 2011;95(3):507–23.

Chapter 4
The Use of Patient Reported Outcome Questionnaires in the Diagnosis of Androgen Deficiency

Gregory J. Lowe, George C. Bailey, and Tracey L. Krupski

Introduction

Testosterone deficiency syndrome (TDS) is a combination of clinical symptoms and a low serum testosterone level. Testosterone deficiency may cause significant morbidity, a decrease in health-related quality of life (HRQoL), or both. Signs of testosterone deficiency can be categorized into sexual side effects, constitutional, cognitive, and physiologic. Classically, urologists think of "sexual" implications such as low libido, gynecomastia, sexual dysfunction, and impaired fertility. However, examples of the constitutional effects include loss of muscle mass, increased abdominal fat deposition, poor work-out recovery, fatigue, hot flashes, and loss of body hair. More subtle changes often not perceived by the patient or physician includes mood instability, reduced ability to concentrate, and memory deficits. Increasingly, the implications of testosterone deficiency on cardiovascular and bone health are being recognized. Treatment is directed at symptomatic improvement and health maintenance. Many testosterone replacements normalize serum levels of testosterone; however, symptomatic improvement is less consistent.

Symptoms experienced in patients with TDS are not unique to this disorder. Many of the symptoms are multi-factorial and can be associated with other common medical conditions. Other symptoms, such as low libido and fertility impairment, are more specific to testosterone deficiency, but not exclusive to this disease.

G.J. Lowe, MD
Department of Urology, Ohio State University Medical Center, Columbus, OH, USA

G.C. Bailey
Department of Urology, Mayo Clinic, Rochester, MN, USA

T.L. Krupski, MD, MPH (✉)
Department of Urology, University of Virginia, P.O. Box 800422,
Charlottesville, VA 22908, USA
e-mail: tlk6f@virginia.edu

W.J.G. Hellstrom (ed.), *Androgen Deficiency and Testosterone Replacement: Current Controversies and Strategies*, Current Clinical Urology,
DOI 10.1007/978-1-62703-179-0_4,

Screening HRQoL questionnaires have been developed to assist the physician in determining which patient needs further evaluation for testosterone deficiency, and to assess symptomatic improvement after treatment has been initiated. In this chapter, we review these screening devices and describe the test characteristics of each.

Health-Related Quality of Life Measures

Prior to a review of the specific questionnaires used to measure hypogonadal symptoms, it is important to understand the basis for developing these questionnaires. HRQoL questionnaires encompass an individual's perception of his own health and functional status. HRQoL researchers term these health domains. A disease process affecting multiple systems such as diabetes will diminish overall functioning and impact general HRQoL. Specific domains vary between questionnaires, but often include physical, emotional, and social domains. However, more subtle disease processes such as prostate cancer diagnosis and treatment will not change general quality of life but do negatively impact disease-specific HRQoL such as continence. This led to the development of disease-specific instruments directed at specific function, such as questions regarding urinary, gastrointestinal, and sexual function. The development process for an instrument includes vigorous testing of psychometric properties focusing on validity and reliability [1]. Validation ensures the instrument truly measures the health aspect one is intending to capture. In addition, convergent validity refers to how closely the new measure correlates with known measures for the same health process. The questionnaire must be reliable and consistent in measuring the intended variables for both the domain and the individual patient. Internal consistency refers to the concept that the items making up each domain all measure the different aspects of that same concept. Test retest reliability means that the same patient (if no changes in health condition occur) should score the same on the instrument if administered several times. Finally, the questionnaire should be responsive. Responsiveness refers to the idea that the instrument needs to change as the health condition changes. For example, an instrument measuring depression should document improvement after initiating an antidepressant. For the best application of HRQoL measures, a baseline value should be documented with subsequent longitudinal follow-up. When available, it is recommended that a neutral third party collect the information. A glossary of terms is provided in Table 4.1.

General Health-Related Quality of Life Measures

The most commonly utilized nonspecific, multidimensional measure is the RAND Medical Outcomes Short Form-36 (SF-36). This questionnaire is made up of several domains including: global health perception, mental health, energy and vitality, pain,

Table 4.1 Definitions of terms

Reliability: a description of how well repeated measurements of the same parameter, with the same instrument, agree with one another
Inter-rater reliability: a description of how well measurements of the same parameter, with the same instrument, by different evaluators, agree with one another
Internal Consistency: a description of how well individual components of an instrument measure the same parameter
Responsiveness: a description of how well an instrument detects change in a parameter
Validity: a description of how well an instrument measures the parameter it is intended to measure
Face validity: a description of how intuitive an instrument's validity appears
Convergent validity: a description of how well an instrument's measurements agree with the measurements of other instruments designed to measure the same parameter
Construct validity: a description of how well an instrument's measurements correlate with parameters known to be associated with the parameter in question
Discriminant validity: a description of how well an instrument's measurements deviate from parameters known not to be associated with the parameter in question

Table 4.2 General HRQoL measures used to assess testosterone deficiency

Author, year	Title	Domains measured
Ware, 1992	SF-36	Constitutional, cognitive
Ware, 1996	SF-12	Constitutional, cognitive, physiologic
Dupuy, 1984	Psychological general well-being index	Constitutional, cognitive, physiologic
Bergner, 1981	Sickness Impact Profile	Constitutional, cognitive
Edicott, 1981	Endicott Quality of Life Enjoyment and Satisfaction Scale	Sexual, constitutional, cognitive, physiologic
Schmidt, 2004	Visual analogue mood Scale	Constitutional, cognitive
Haren, 2005	Likert Scale plus visual analogue Scale	Constitutional, cognitive
Fugl-Meyer, 1997	Life Satisfaction Scale	Sexual, cognitive
Herr, 2000	QL uniscale	Global assessment of quality of health
Park, 2003	Pusan National University Hospital quality of life scoring system	Sexual, constitutional, cognitive, physiologic

social function, physical function, physical role and mental role [2]. This is one example of a nonspecific questionnaire that has been utilized to assess the impact of TDS and its treatment. These generic questionnaires typically do not perform as well as testosterone deficiency specific measures [3]. Occasionally, authors try to utilize a single global question to measure overall quality of life. A specific example of this type of measure is the Quality of Life uniscale (QL uniscale), also known as the Spitzer uniscale QL [4]. One-item instruments are rarely useful as they are not accurate enough to capture the gradual changes most patient experience over time. Other questionnaires have multiple domains to assess quality of life. The most frequently used general instruments are summarized in Table 4.2. Alternative general HRQoL multidimensional questionnaires include the Rand Medical Outcomes Short form -12 (SF-12 v1 and SF-12 v2), Sickness Impact Profile, Endicott Quality of Life

Enjoyment and Satisfaction Scale, Psychological general well-being index, Pusan National University Hospital Quality of life scoring Scale, Life Satisfaction Scale, Visual analogue mood Scale, and the Likert Scale plus visual analogue Scale. While these may capture the constitutional or cognitive effects of hypogonadism, only three of these Scales include a sexual function domain: Pusan National University Hospital Quality of life scoring system, Life Satisfaction Scale, and Endicott Quality of Life Enjoyment and Satisfaction Scale. However, none of these scales contain both the energy/vitality and sexual function domains.

Langham et al. assessed the use of nonspecific tools for following response to treatment of hypogonadism [3]. Of the 14 intervention studies, seven showed a positive impact of treatment in a specific domain over time or between groups. Five studies showed no change and two studies revealed a negative impact of treatment. The generic tools performed poorly when evaluated for clinical face validity. They only performed better than specific tools in regards to incorporating a global rating separate from other domains.

General HRQoL questionnaires are useful to compare across populations or between different disease states. They can also augment the utility of disease-specific measures to make the findings applicable across a wider patient population.

Testosterone Deficiency Specific Questionnaires

Disease-specific questionnaires for TDS are available and improve on validity, relevance, and responsiveness compared to general HRQoL instruments. The instruments and the domains they address are found in Table 4.3. Unfortunately, even these specific questionnaires are imperfect to screen for testosterone deficiency. In one review of several specific questionnaires, there was evidence of greater ability to detect a change after treatment compared to the general HRQoL instruments [3]. Additionally, it was felt that all of these questionnaires were clinically relevant to assess for symptoms related to hypogonadism.

Table 4.3 Testosterone deficiency specific questionnaires

Author, year	Title	Domains measured
Heinemann, 1999	Aging Males' Symptoms Scale	Sexual, constitutional, cognitive, physiologic
Morley, 2000	Androgen Deficiency in Aging Males	Sexual, constitutional, cognitive
Smith, 2000	Massachusetts Male Aging Study	Sexual, constitutional, cognitive, physiologic
Wiltink, 2009	Hypogonadism-Related Symptom Scale	Sexual, constitutional, cognitive
Corona, 2009	Androtest	Sexual, constitutional, cognitive
McMillan, 2003	Age-Related Hormone Deficiency-Dependent Quality of Life Questionnaire	Sexual, constitutional, cognitive

Aging Males' Symptoms Scale

The most extensively studied specific questionnaire is the Aging Males' Symptom (AMS) Scale. This Scale is a 17 item questionnaire used to assess patient complaints in three domains: psychological (five items), somatovegetative (seven items), and sexual (five items). A Likert Scale is used for each question with 1 indicating no complaint and 5 noting severe bother. A total score is calculated with ranges available to characterize overall symptoms and likelihood of androgen deficiency. The total score is graded as: 17–26 points representing "no/little symptoms," 27–36 points representing "mild symptoms," 37–49 points representing "moderate symptoms," and 50–85 points representing "severe symptoms."

The AMS Scale was originally developed in 1999, in Germany [5]. It was created to address the idea that men may experience events analogous to female menopause. Over 200 variables were assessed using factorial analysis to identify complaints and domains specifically related to aging. The goals of the symptom Scale development were to assess symptoms of aging (independent from disease-related symptoms) in groups of men under different conditions, to evaluate the severity of symptoms over time, and to measure changes pre- and post-androgen replacement. Initial testing was performed on men over 40 years of age. At its inception, the AMS Scale was not intended to be a screening tool for androgen deficiency, although it was intended to measure treatment effect. Normal standardized scores have only been published for German and Japanese patients. The Scale has been used internationally and translated into many other languages including English, Dutch, French, Spanish, Portuguese, Italian, Swedish, Korean, and Indonesian.

The AMS Scale has good reliability; with a Cronbach's Alpha measure of internal consistency between 0.7 and 0.9 across countries, time periods, and domain subscores [6]. This was determined from a meta-analysis of studies performed in Germany, the United Kingdom, Spain, Portugal, Italy, France, Sweden, Thailand, and Korea. Small patient numbers assessed in some countries limits the reliability determination. Validity was assessed as well, revealing the ability to discriminate between degrees of hypogonadism [6].

Several studies have used the AMS to evaluate symptom response to androgen replacement. The AMS scores for patients mirror those results determined by physician evaluation of treatment efficacy [7]. In a study of nearly 900 men, the average total score before treatment was 45.3 ± 13.2 compared to 29.9 ± 9.1 after treatment, for men above 50 years of age. Similar results were seen for men less than 50 years of age. Prior to therapy, only 5.7% of patients reported "no/little symptom" scores, but this increased to 41.8% after treatment with injectable testosterone enanthate. Unfortunately, the only randomized, double-blind, placebo controlled study to assess testosterone replacement therapy in men with laboratory proven androgen deficiency revealed no relationship between the AMS score and free testosterone level [8]. This study also did not find any statistically significant difference between placebo and treatment group AMS scores after 6 months of testosterone replacement.

Additional studies evaluating the relationship between total AMS score and serum testosterone levels document variable results. Several abstracts and articles have shown a positive correlation between total or free testosterone and the total AMS score [9, 10]. While AMS score alone was shown to be predictive of testosterone levels, the addition of body mass index (BMI) and age further improved predictive ability. In contrast, others studies were unable to reveal the same finding and suggest testosterone is not related to total AMS score [11–13].

A number of factors exist which potentially explain these conflicting results. The method of hormone level determination and patient demographics may influence study results [14]. The AMS Scale has also been related to lower household income and major depressive disorder [15]. Although the Scale is criticized for being too long and requiring too much time to complete [16], it has shown the ability to be used over the phone or in person making it useful for research studies. Currently, the AMS Scale should be viewed as an ancillary tool to track symptom severity and change in symptoms over time in patients. It has limited utility in screening patients for androgen deficiency, but this weakness is mitigated when patient AMS score is combined with BMI and age (Appendix 1).

Androgen Deficiency in Aging Males

The Saint Louis University Androgen Deficiency in Aging Males (ADAM) questionnaire is composed of ten "yes or no" questions regarding patient symptoms related to sexual function, mood, and energy level. A positive score answers "yes" to questions 1 or 7; otherwise a "yes" answer is needed on any three other questions [14]. The Scale was initially developed by identifying ten symptoms common to patients with low bioavailable testosterone. The ADAM questionnaire has been translated and validated in Chinese and Arabic. Several other nonvalidated translations are also available.

Initial results of the ADAM questionnaire reported an 88% sensitivity and 60% specificity [17]. Later studies have supported the 80–90% sensitivity but reported lower specificity of 19–36% [18–22]. One study found age and diabetes mellitus to be correlated with a positive ADAM questionnaire independent of a low bioavailable testosterone level [18]. The internal reliability is adequate with a Cronbach's alpha of 0.71–0.74 [19, 20].

Similarly to the AMS, correlating the ADAM questionnaire with testosterone levels is problematic. Results are conflicting with several publications documenting the ADAM questionnaire to be a sensitive indicator of low bioavailable testosterone [18–22]. The two studies have noted no correlation between serum bioavailable testosterone and a positive ADAM questionnaire [23, 24]. The first question ("Do you have a decrease in libido?") has been shown to be more specific than other individual questions or the total score. One study supported the idea that men with a free testosterone level less than 70 ng/dL will have at least one symptom from each domain (sexual, energy, mood) [22].

There have been several modifications of the ADAM questionnaire. The most frequent modification is to calculate a cumulative score based on the total number of "yes" responses, with each "yes" response representing one point. This has shown mixed results with one positive and one negative study [22, 23]. A short version of the questionnaire consisting of six items has been shown to increase specificity in a Chinese patient population [20]. A more recent variation of the ADAM questionnaire applies a Likert Scale of 1–5 for each question [25]. Unlike the AMS, a score of 1 indicates severe symptoms, and a score of 5 indicates minimal to no trouble with each item. Therefore, a score of 10 represents the most symptomatic, while a score of 50 is least symptomatic. This quantitative version of the ADAM questionnaire was tested in a population of 57 men scheduled to undergo radical prostatectomy. It was found to correlate positively with both serum testosterone values and the sexual health inventory for men questionnaire.

Overall, the ADAM questionnaire has been shown to have high sensitivity, but low specificity for diagnosing androgen deficiency. It is useful to rule out a patient with a negative questionnaire, but does not reliably rule in a patient with a positive result. Therefore, it should not be used as a surrogate for a serum bioavailable testosterone in a patient that a physician is suspicious for androgen deficiency (Appendix 2).

Massachusetts Male Aging Study/Smith Questionnaire

The Massachusetts Male Aging Study (MMAS) is an eight-item questionnaire interrogating the presence of risk factors for hypogonadism. This instrument grew out of a community-based observational study randomly sampling men aged 40–70 [26]. Blood samples, physiological measures, socio-demographic variables, psychological indexes, and information on health status, medications, smoking, and lifestyle were collected. Risk factors evaluated were smoking, asthma, headaches, sleep, diabetes, age, and desire to be in charge of others. Each item response is weighted according to its strength of association with low testosterone providing a score between 0 and 15. A score of greater than five is suggestive of a possible testosterone deficiency, and half of men with scores above 10 will be hypogonadal. An initial evaluation based on these thresholds revealed a sensitivity of 71% and specificity of 53% [27]. This questionnaire was designed to be intuitive enough that patients could complete the questionnaire, interpret the results at home, and decide on their own if further testing was needed (Appendix 3).

The authors of the MMAS anticipated that smoking would be associated with testosterone deficiency. As the questionnaire was constructed, smoking status turned out to be one of the eight most significant predictors of testosterone deficiency. Other investigators have shown an opposite effect; smoking was not related to androgen levels [28]. In fact, the history of conflicting results seen among those who have explored the relationship between smoking and testosterone deficiency may help account for the mixed results observed in attempts to use the MMAS as a screening tool.

Hypogonadism Related Symptoms Scale

The Hypogonadism-Related Symptoms (HRS) Scale was developed in 2009 in an effort to find a screening tool with improved specificity over other available scales [29]. Multi-factorial analysis of items from multiple, previously validated general HRQoL questionnaires were evaluated for correlation with testosterone levels. This evaluation found 19 items, covering five domains, which were related to serum testosterone levels. These domains included reduced physical activity, dissatisfaction with sexual function, a negative self-concept of physical fitness, reduced sexual desire, and hot flashes. In developing this Scale, 263 men were asked to complete an ADAM and AMS questionnaire as well for construct validity. This revealed a weak correlation between total testosterone and the ADAM questionnaire, and no correlation between AMS and total testosterone. Free testosterone did not correlate with either AMS or ADAM questionnaires. The HRS Scale total score correlated weakly with free testosterone and moderately well with total testosterone levels. The internal reliability was excellent with a Cronbach alpha of 0.87.

Creating a Scale based on refining items already known to be associated with testosterone level is a reasonable approach. One would expect this instrument should perform better in screening and assessment of symptom response to hormone replacement therapy. However, as this is such a recently developed instrument, few authors have used it in the clinical arena. Further studies are necessary to establish the validity of the HRS.

Androtest

The Androtest was designed to be a structured interview with specific guidelines for quantifying patient responses [30]. The questions selected are typical sexual health evaluation queries. Twelve questions are scored with a total score range from 0 to 32. Questions are weighted and ranged from 0 to 3 for each question. These questions cover several domains including sexual and ejaculation function and medical conditions associated with hypogonadism. The Androtest was explicitly designed to screen men presenting with specific sexual dysfunction complaints. A score of greater than eight has a sensitivity of 68% and specificity of 65% for detection of a total testosterone less than 300 ng/dL [30].

The Androtest has been used to assess the prevalence of hypogonadism-associated symptoms in patients on statin therapy and in patients presenting for sexual dysfunction [31, 32]. Statin use was correlated with lower serum testosterone levels and higher Androtest scores; however, no direct comparison of Androtest score was made with serum testosterone level. No study could be found using Androtest pre- and post-androgen replacement to evaluate for response to testosterone treatment.

One benefit of the Androtest is also a limitation. A structured interview allows the physician to explain questions the patient may not understand, but may lead to inadvertent bias. The purported specificity of the Androtest of 68% does suggest greater specificity for detecting low testosterone symptoms than any of the aforementioned instruments. However, few publications exist utilizing this instrument and it has not been validated or studied outside the Italian language.

Age-Related Hormone Deficiency-Dependent Quality of Life Questionnaire

The Age-Related Hormone Deficiency-Dependent Quality of Life (A-RHDQoL) questionnaire is a 21 item measure of a subject's perceived impact of testosterone decline [33]. Unique to this questionnaire is the ability for patients to rate the importance of each aspect on their life. The questions also address what changes a patient would expect if they did not have a hormone deficiency. The domains covered include family life, social life, work life, sex life, physical capabilities, appearance, stamina, sleep, pain, confidence, motivation, fertility, concentration, and ability to perform household tasks. Each domain question is scored from −3 to 3 and each item is also assigned an importance score of zero to three. The domain score is multiplied by the importance score for a weighted score per question. The bipolarity of the Scale allows patients to report positive impacts of age-related hormone decline. All questions completed (not marked NA) are totalled and divided by the total number of completed items creating an Average Weighted Impact score. For completeness, the final item in this questionnaire is an open-ended free comments section that asks patients to list any other ways hormone deficiency impacts their HRQoL.

The A-RHDQoL has high reliability with a Cronbach's alpha of 0.935 [33]. Question one, measuring present quality of life, is significantly correlated with total testosterone levels. There were no other significant correlations between quality of life scores and hormone levels. In initial testing, 14% of patients reported difficulty answering the questions due to a lack of knowledge of the effects of declining hormone levels on various aspects of life. No validation has been performed on this questionnaire.

Direct Comparisons

Several studies have evaluated the agreement between instruments. Nearly all of these have included the AMS Scale, the ADAM questionnaire and the MMAS. In 2004, 145 German males completed the AMS, ADAM, and MMAS questionnaires over the telephone [34]. The sequence of the questionnaires was varied and interviews were completed by a marketing institute. In this patient population, the AMS Scale predicted "no/little complaints" in 82%, corresponding to an ADAM negative questionnaire in 43% and an MMAS score of "hypogonadism unlikely" in 46%.

Table 4.4 Sensitivity and specificity of disease-specific instruments with the gold standard

	ADAM		MMAS		AMS	
	Sensitivity	Specificity	Sensitivity	Specificity	Sensitivity	Specificity
Low testosterone						
Bioavailable testosterone <70 ng/dL (%)	97	30	60	59	83	39

Negative results with ADAM were almost exclusively found in men with AMS "no/little symptoms" (97%). Using the AMS score to predict the ADAM score resulted in a positive predictive value of 92% and negative predictive value of 50%. The sensitivity was calculated to be 29% but specificity was 97%. The AMS Scale's ability to predict the MMAS outcome was found to be 65% and 49% respectively for positive and negative predictive value. Sensitivity was again low at 22% and specificity was 87%. The ability of the MMAS outcome to predict the ADAM findings were similar to those shown by the AMS Scale. The authors of this study concluded that no instrument demonstrated diagnostic superiority.

A similar study used the same three questionnaires, but compared these with bioavailable testosterone levels as the "gold standard." [35] One hundred and forty-eight men aged 23–80 years completed the questionnaires. The ADAM sensitivity was 97% and specificity was 30% for identifying low bioavailable testosterone levels. The MMAS performed better on specificity (59%) but lower in sensitivity (60%). The AMS Scale was found to have 83% sensitivity and 39% specificity. The authors concluded the ADAM and AMS questionnaires are superior to the MMAS as screening measures for hypogonadism. See Table 4.4 for a comparison of each instrument with bioavailable testosterone.

Two more recent publications evaluated AMS and ADAM questionnaires in determining response to testosterone replacement therapy [8, 36]. One study was a 6-month randomized, double-blind placebo controlled trial of 237 men receiving testosterone undecanoate or placebo. The authors found no significant association between total testosterone and questionnaire outcomes. After testosterone supplementation, there were no effects on the questionnaire outcomes when compared with placebo. Both questionnaires were associated with age however. Older individuals had worse scores on each of the instruments. The second study included 56 patients who underwent 6 months of testosterone replacement therapy, and completed the AMS and ADAM Scales before and after treatment. Prior to therapy, there was no significant correlation observed between free testosterone level and ADAM questionnaire or total AMS score. There was also no correlation noted to individual parameter scores on the AMS Scale. Following testosterone replacement, significant improvement was seen in free testosterone levels, AMS total and psychological parameter scores and the ADAM questionnaire outcome. Taken together, these studies suggest poor screening ability for both the AMS and ADAM questionnaires. The questionnaires may be useful in treatment follow-up but this needs to be further established prior to their use to track patient outcomes.

Recommendations

The most recent guidelines from a consensus group representing the International Society of Andrology, International Society for the Study of Aging Male, European Association of Urology, European Academy of Andrology, and the American Society of Andrology recommend the diagnosis of hypogonadism require the presence of symptoms and signs suggestive of testosterone deficiency corroborated with a low serum testosterone level [37]. *Questionnaires are not recommended for the diagnosis of TDS due to low specificity.*

Conclusion

TDS symptoms are varied and nonspecific to the hypogonadal state, necessitating a physician's clinical suspicion for the presence of low serum testosterone. This chapter reviewed the most commonly utilized questionnaires currently available to assist in screening and monitoring therapy. Unfortunately, many limitations of these questionnaires exist, and the scales are not specific enough to reliably rule in or exclude, TDS. Common comorbidities such as aging or depression show significant symptom overlap with androgen deficiency. Current recommendations do not support the use of available questionnaires as the sole instrument for screening or diagnosis. Understanding the inherent limitations allows the clinician to use the AMS, ADAM, MMAS, Androtest, HRS, or A-HRDQoL as an adjunct measure in diagnosing and treating the androgen deficient male.

References

1. Foster DS, Kuhn J. How to measure survey reliability and validity. Thousand oaks: SAGE Publications; 1995.
2. Ware Jr JE, Sherbourne CD. The MOS, 36-item short-form health survey (SF-36). I. Conceptual framework and item selection. Med Care. 1992;30:473–83.
3. Langham S, Maggi M, Schulman C, et al. Health-related quality of life instruments in studies of adult men with testosterone deficiency syndrome: a critical assessment. J Sex Med. 2008;5:2842–52.
4. Spitzer WO, Dobson AL, Hall J, et al. Measuring the quality of life if cancer patients: a concise QL index for use by physicians. J Chronic Dis. 1981;34:585.
5. Heinemann LA, Zimmermann T, Vermeulen A, Thiel C. A new 'Aging Male's Symptom' rating scale. Aging Male. 1999;2:105–14.
6. Daig I, Heinemann LA, Kim S, et al. The Aging Males' Symptoms (AMS) scale: review of its methodological characteristics. Health Qual Life Outcomes. 2003;1:77.
7. Moore C, Huebler D, Zimmermann T, et al. The Aging Males' Symptoms scale (AMS) as outcome measure for treatment of androgen deficiency. Eur Urol. 2004;46:80–7.
8. Emmelot-Vonk M, Verhaar H, Nakhai-Pour H, et al. Low testosterone concentrations and the symptoms of testosterone deficiency according to the Androgen Deficiency in Ageing Males

(ADAM) and Ageing Males' Symptoms rating scale (AMS) questionnaires. Clin Endocrinol (Oxf). 2011;74:488–94.
9. Kratzik C, Reiter W, Riedl A, et al. Hormone profiles, body mass index and aging male symptoms: results of the Androx Vienne Municipality study. Aging Male. 2004;7:188–96.
10. Jankowska E, Lopuszanska M, Szklarska A, Medras M. Hormonal determinants of andropausal symptoms in Polish men. Aging Male. 2004;7:21.
11. T'Sjoen G, Feyen E, De Kuyper P, et al. Self-referred patients in an aging male clinic: much more than androgen deficiency alone. Aging Male. 2003;6:157–65.
12. Kratzik C, Heinemann LA, Saad F, et al. Composite screener for androgen deficiency related to the Aging Males' Symptoms scale. Aging Male. 2005;8:157–61.
13. Miwa Y, Kaneda T, Yokoyama O. Correlation between the Aging Males' Symptoms scale and sex steroids, gonadotropins, dehydroepiandrosterone sulfate, and growth hormone levels in ambulatory men. J Sex Med. 2006;3:723–6.
14. Raynaud J, Tichet J, Born C, et al. Aging male questionnaire in normal and complaining men. J Sex Med. 2008;5:2703–12.
15. Yoshida N, Kumano H, Kuboki T. Does the Aging Males' Symptoms scale assess major depressive disorder: a pilot study. Maturitas. 2006;53:171–5.
16. O'Leary M. Development of an index to evaluate symptoms in men with androgen deficiency. Rev Urol. 2003;5 Suppl 1:S11–5.
17. Morley J, Charlton E, Patrick P, et al. Validation of a screening questionnaire for androgen deficiency in aging males. Metabolism. 2000;49:1239–42.
18. Martinez-Jabaloyas J, Queipo-Zaragoza A, Rodriguez-Navarro R, et al. Relationship between the Saint Louis University ADAM questionnaire and sexual hormonal levels in a male outpatient population over 50 years of age. Eur Urol. 2007;52:1760–7.
19. Rabah D, Arafa M. Validation of an Arabic ADAM questionnaire for androgen deficiency screening in the Arab community. Aging Male. 2009;12:95–9.
20. Chu L, Tam S, Kung A, et al. A short version of the ADAM questionnaire for androgen deficiency in Chinese men. J Gerontol. 2008;63:426–31.
21. Blumel J, Chedraui P, Gili S, et al. Is the Androgen Deficiency of Aging Men (ADAM) questionnaire useful for the screening of partial androgenic deficiency of aging men? Maturitas. 2009;63:365–8.
22. Tancredi A, Reginster J, Schleich F, et al. Interest of the androgen deficiency in aging males (ADAM) questionnaire for the identification of hypogonadism in elderly community-dwelling male volunteers. Eur J Endocrinol. 2004;151:355–60.
23. Spetz A, Palmefors L, Skobe R, et al. Testosterone correlated to symptoms of partial androgen deficiency in aging men (PADAM) in an elderly Swedish population. Menopause. 2007;14:999–1005.
24. Hanus M, Matouskova M, Starka L, Hill M. Hormonal homeostasis in a group of 216 aging Czech males and correlation with responses to a questionnaire of the University of St Louis. Aging Male. 2006;9:103–10.
25. Mohamed O, Freundlich R, Dakik H, et al. The quantitative ADAM questionnaire: a new tool in quantifying the severity of hypogonadism. Int J Impot Res. 2010;22:20–4.
26. Feldman HA, Goldstein I, Hatzichristou DG, et al. Construction of a surrogate variable for impotence in the Massachusetts Male Aging Study. J Clin Epidemiol. 1994;45:457–67.
27. Smith K, Feldman H, McKinlay J. Construction and field validation of a self-administered screener for testosterone deficiency (hypogonadism) in ageing men. Clin Endocrinol. 2000;53:703–11.
28. Halmenschlager G, Rossetto S, Muller Lara G, et al. Evaluation of the effects of cigarette smoking on testosterone levels in adult men. J Sex Med. 2009;6:1763–72.
29. Wiltink J, Beutel M, Brahler E, et al. Hypogonadism-related symptoms: development and evaluation of an empirically derived self-rating instrument (HRS 'Hypogonadism Related Symptom scale'). Andrologia. 2009;41:297–304.
30. Corona G, Mannucci E, Petrone L, et al. ANDROTEST: a structured interview for the screening of hypogonadism in patients with sexual dysfunction. J Sex Med. 2006;3:706–15.

31. Corona G, Boddi V, Balercia G, et al. The effect of statin therapy on testosterone levels in subjects consulting for erectile dysfunction. J Sex Med. 2010;7:1547–56.
32. Corona G, Mannucci E, Lotti F, et al. Impairment of couple relationship in male patients with sexual dysfunction is associated with overt hypogonadism. J Sex Med. 2009;6:2591–600.
33. McMillan C, Bradley C, Giannoulis M, et al. Preliminary development of a new individualised questionnaire measuring quality of life in older men with age-related hormonal decline: the A-RHDQoL. Health Qual Life Outcomes. 2003;1:51.
34. Heinemann LA, Saad F, Heinemann K, et al. Can results of the Aging Males' Symptoms (AMS) scale predict those of screening scales for androgen deficiency? Aging Male. 2004;7:211–8.
35. Morley J, Perry H, Kevorkian R, et al. Comparison of screening questionnaires for the diagnosis of hypogonadism. Maturitas. 2006;53:424–9.
36. Yamaguchi K, Ishikawa T, Chiba K, et al. Assessment of possible effects of testosterone replacement therapy in men with symptomatic late-onset hypogonadism. Andrologia. 2010; 43:52–6.
37. Wang C, Nieschlag E, Swerdloff R, et al. Investigation, treatment, and monitoring of late-onset hypogonadism in males: ISA, ISSAM, EAU, EAA, and ASA recommendations. Eur J Endocrinol. 2008;159:507–14.

Chapter 5
Testosterone and Its Association with Metabolic and Cardiovascular Disease

Giovanni Corona, Giulia Rastrelli, and Mario Maggi

Introduction

Since the 1970s, epidemiological evidence has clearly documented that, regardless of how life expectancy is measured, women live longer than men. The average life expectancy for women worldwide is 68 years, while for men it is 64 years [1]. Explanations for women's advantageous mortality profile can be classified broadly into three categories: behavioral, sociocultural, and biological.

In developed countries, men live more risky and unhealthy lifestyles, which contribute to increased male mortality. Adult women under the age of 65 visit the doctor more frequently than men, with the widest gender gap between ages 18 and 44, and attributed to medical care associated with reproduction [1]. Men consume alcohol more frequently and in higher amounts than women [2, 3], and are more than twice as likely to die from alcoholic liver disease and cirrhosis [4]. Men are also two to four times more likely to die prematurely from unintentional injury, homicide, and suicide [4]. Men also out-smoke women, although this difference is nonexistent at the youngest ages, which reveals that smoking rates are equal in the youngest age cohorts [5, 6]. The rising rate of smoking among men, relative to women, accounted for 75% of the increase in gender mortality between 1910 and

G. Corona, MD, PhD
Andrology and Sexual Medicine Unit, Department of Clinical Physiopathology, University of Florence, Florence, Italy

Endocrinology Unit, Maggiore-Bellaria Hospital, Bologna, Italy

G. Rastrelli, MD • M. Maggi, MD (✉)
Andrology and Sexual Medicine Unit, Department of Clinical Physiopathology, University of Florence, Florence, Italy
e-mail: m.maggi@dfc.unifi.it

W.J.G. Hellstrom (ed.), *Androgen Deficiency and Testosterone Replacement: Current Controversies and Strategies*, Current Clinical Urology,
DOI 10.1007/978-1-62703-179-0_5, © Springer Science+Business Media, LLC 2013

1962 [1, 7]. Current estimates suggest that tobacco abuse accounts for 25% of the overall sex mortality difference, and is much higher when cancer and respiratory disease mortality are considered separately. These lifestyle behaviors result in worsening health and increased likelihood of hospital admission and ultimately death.

Sociocultural factors play another essential role in explaining gender differences in mortality. Differences exist in what is expected of men and women and how one is taught to behave. For example, many cultures encourage or condone a man's heavy drinking, but discourage it in women. In many cultures, women are not encouraged to work outside the home, while men are expected to be part of the labor force. Because women work less in a physical sense than men, they escape the toils of labor, and their health deteriorates less rapidly [1].

From a biological standpoint, women are offered some protection against mortality. Specifically, male sex hormones may help explain male behavioral patterns related to overall health. In particular, evidence published over the last few decades has pointed out that T may account, in part, for the life expectancy in males, due to an increased incidence of cardiovascular disease (CVD), creating the myth that T supplementation is dangerous, though recent evidence reveals low T rather high T is associated with a higher male morbidity and mortality [7–9].

Epidemiological Association Between Testosterone and Cardiovascular Risk Factors

Obesity

A decline in both sex-hormone-binding-globulin (SHBG)-bound and -unbound testosterone is associated with obesity [7, 9–12] (see also Fig. 5.1). An important observation is that luteinizing hormone (LH) levels do not significantly rise in obese subjects to compensate for obesity-induced testosterone decline (see Fig. 5.1), suggesting the existence of functional hypogonadotropic hypogonadism. Accordingly, previous studies performed in morbidly obese men indicated that LH levels and pulse amplitude were attenuated as compared to normal weight controls [13, 14]. In an experimental rabbit model of high fat diet (HFD)-induced metabolic syndrome (MetS), HFD induced overt hypogonadotropic hypogonadism, associated with reduced gonadotropin releasing hormone (GnRH) immunopositivity in the arcuate nucleus of the hypothalamus.

Type 2 Diabetes Mellitus

Cross-sectional studies have demonstrated an association between low T and type 2 diabetes mellitus (T2DM) since the early 1980s [7, 15–17]. Recently, Anderson

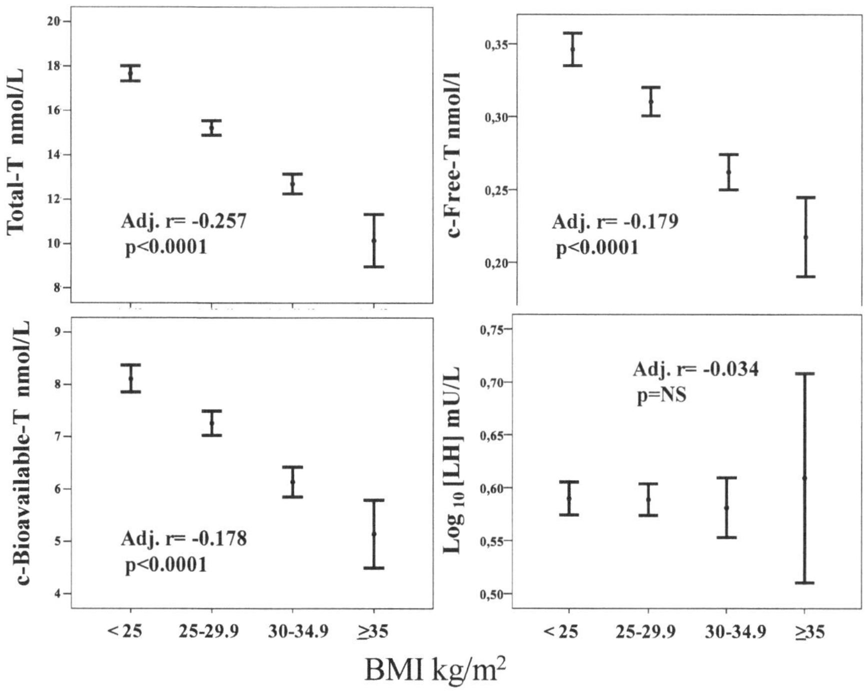

Fig. 5.1 (**a–d**) Total, calculated free, calculated bioavailable testosterone (T), and LH (log. transformed) levels as a function of obesity classes. Data are expressed as mean (95% confidence interval). *BMI* body mass index. Insets indicate the age and chronic disease score index adjusted data. Data were obtained in a consecutive series of 1,715 subjects attending our Unit seeking medical care for sexual dysfunction between 2000 and 2011

et al. [18], in a large series of 6,457 male patients aged 18–80 years with diabetes, demonstrated that 4.4% of them were frankly hypogonadal with a serum total testosterone of less than 8.0 nmol/L (~230 ng/dL). For borderline hypogonadism (serum total testosterone 8–11.99 nmol/L (~230–345 ng/dL)) the proportion of T2DM in men rose to 32.1%. In line with these data, two independent meta-analyses [19, 20] indicated that total T was significantly lower in men with T2DM. As a corollary, longitudinal data have demonstrated that low T predicts incident diabetes [20]. Inversely T2DM, obesity, and MetS predict forthcoming hypogonadism [19]. Although the direction and the dimension of the association between low T and T2DM have to be further investigated, current recommendations of most important academic societies [21, 22] consider T2DM as a risk factor for male hypogonadism. The specific mechanisms involved in the pathogenesis of T2DM-associated hypogonadism are complex and have not been fully elucidated. In line with the data observed in obesity, Dhindsa et al. [23] reported that in T2DM, androgen deficiency is commonly associated with an impaired gonadotropin response, leading to hypogonadotropic hypogonadism.

Hypertension

Epidemiological evidence regarding the association between T levels and hypertension is conflicting. Androgens have been attributed to the development and severity of hypertension in some genetic and nongenetic rat models [24, 25]. In rats that are hypertensive but not obese, androgens increase their blood pressure [26, 27]. However, contrasting findings have been reported in epidemiological trials. In particular, some studies have shown reduced androgen levels in subjects with essential hypertension, as compared to normotensive subjects [28, 29] while other investigators could not replicate this result [30]. In subjects with sexual dysfunction and MetS studies did not demonstrate any significant contribution of the hypertensive condition to male hypogonadism [31]. Conversely, there was an inverse relationship between SHBG-bound and unbound testosterone levels and pulse pressure, a surrogate marker of arterial stiffness [32]. Similar results were previously reported by Fogari et al. [33] in a consecutive series of 356 nondiabetic, nonsmoking, non-obese men aged 60–80 years and untreated for hypertension. These results suggest that in elderly men with systolic hypertension the reduced plasma levels of T may contribute to increased arterial stiffness. The mechanisms underlying these T effects remain unclear. Recent studies suggest that insulin resistance may be the common pathogenetic link between low T and arterial stiffness (see below [7]).

Dyslipidemia

Evidence has demonstrated inconsistent associations between reduced T levels and abnormal lipid profiles. Initial studies demonstrated increased total cholesterol, low density lipoprotein (LDL), and triglyceride (TG) levels among hypogonadal patients with improvements noted following supplementationW [34, 35]. With long-term androgen deprivation (12 months) among patients with prostate cancer, increases in total LDL, high density lipoprotein (HDL), and TG have been noted, while shorter term studies have demonstrated conflicting results with LDL and increases in HDL and total cholesterol [36–38]. Further studies have demonstrated no differences in HDL, LDL, total cholesterol, or TG levels [39–41].

Metabolic Syndrome

A large body of evidence suggests that male hypogonadism is frequently associated with MetS [9, 12, 31, 42]. Meta-analyses of available epidemiological reports revealed that MetS is significantly associated with an overall lower total T [42, 43], the difference being more evident in studies conducted in subjects with ED

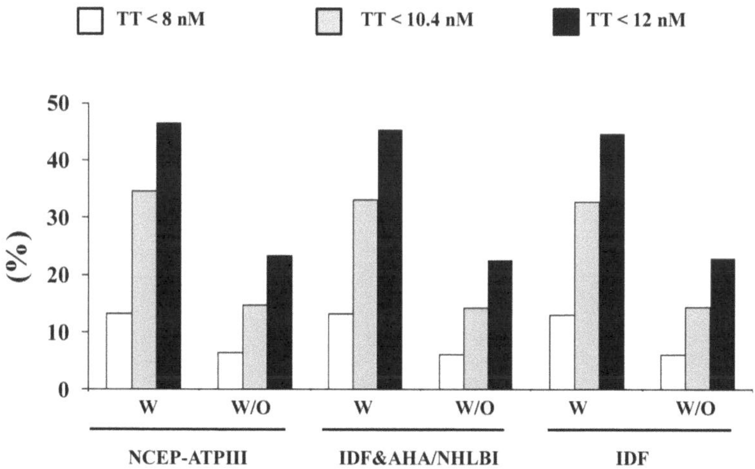

Fig. 5.2 Prevalence of hypogonadism according to different total testosterone thresholds in patients with (W) and without (W/O) different metabolic syndrome (MetS) definitions. Data are derived from a consecutive series of 2,481 patients attending our unit for sexual dysfunction between 2000 and 2011. NCEP-ATPIII National Cholesterol Education Program-Third Adult Treatment Panel, IDF International Diabetes Federation, AHA/NHLBI American Heart Association/National Heart, Lung and Blood Institute

(−3.51[−4.48; −2.53]) than in those without ED (−2.60[−3.15; −2.06]) [42]. In subjects consulting our outpatient clinic for sexual dysfunction the proportion of subjects identified as having hypogonadism and MetS ranges from about 13% to more than 46%, according to various biochemical thresholds of T deficiency (TD) (see Fig. 5.2). Interestingly, whatever the definition of MetS or TDs (total T lower than 8, 10.4, or 12 nmol/L) is used the relative fraction of hypogonadal subjects identified by MetS does not change. Recent data from the European Male Aging Study (EMAS), a population-based survey of more than 3,400 men enrolled across eight European centers, demonstrated that among many symptoms, sexual issues (ED, reduced libido, and decreased spontaneous erections) were the most specific for TDs [44]. By applying these EMAS-derived criteria (simultaneous presence of three sexual symptoms and T<11 nmol/L), there was a significant association with MetS (see Fig. 5.3).

Pathogenesis of MetS-Related Hypogonadism

The vicious cycle between MetS and hypogonadism is summarized in Fig. 5.4. The androgen receptor is highly expressed in visceral fat and negatively regulates the differentiation of preadipocytes into mature adipocytes [45]. In addition, androgens decrease fat mass by regulating the differentiation of mesenchymal stem cells to adipocytes [46]. In humans, T acutely inhibits oleic acid uptake in omental and

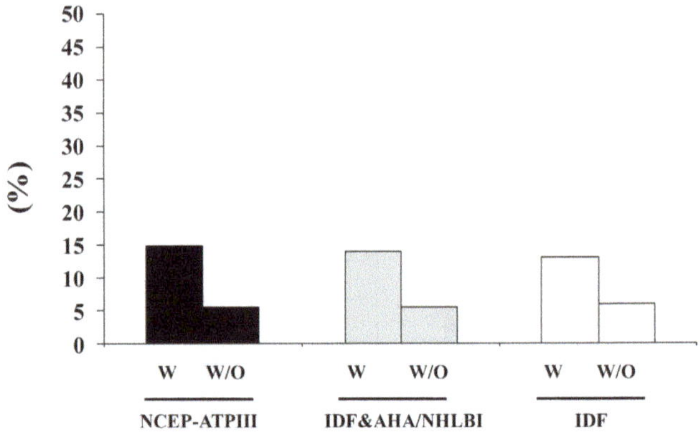

Fig. 5.3 Prevalence of hypogonadism (according to EMAS hypogonadism definition) in patients with (w) and without (w/o) MetS. Data are derived from a consecutive series of 2,481 patients attending our unit for sexual dysfunction between 2000 and 2011. *NCEP-ATPIII* National Cholesterol Education Program-Third Adult Treatment Panel; *IDF* International Diabetes Federation; *AHA/NHLBI* American Heart Association/National Heart, Lung and Blood Institute

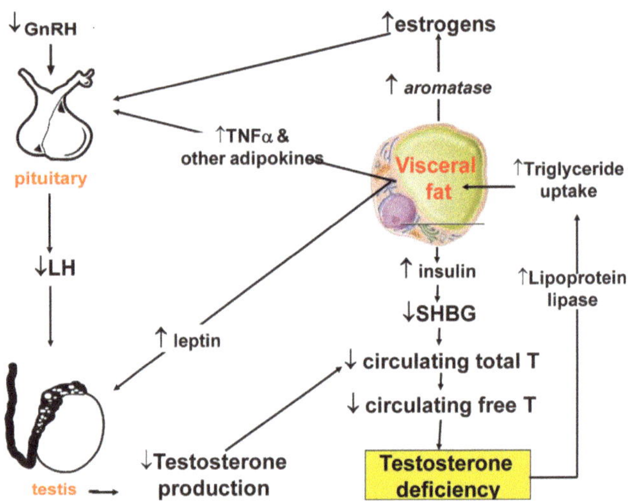

Fig. 5.4 Proposed interactions between increased visceral fat and hypogonadism. *TNFα* tumor necrosis factor α; *T* testosterone (adapted from Buvat J, Maggi M, Gooren L, et al. Endocrine Aspects of Male Sexual Dysfunctions. Journal of Sexual Medicine 2010;7(4):1627-1656, with permission from John Wiley & Sons, Inc.)

retroperitoneal tissues acting through lipoprotein lipase, but not subcutaneous, adipose tissue [47]. A mixed (primary and secondary) Hypogonadal state develops, with the adverse consequence of obesity. Which molecules are responsible for inhibiting the compensatory increase of GnRH/LH in obesity-induced T decline is still unknown, but reasonable candidates are estrogens, insulin, leptin, TNFα, or other adipokines [7, 9, 10, 12, 48].

Hypogonadism can be observed in animal models of insulin-deficient diabetes suggesting a possible role of insulin in regulating hypothalamus GnRH synthesis and release [49–51]. Accordingly, patients with T2DM and MetS, who are insulin resistant, showed a higher prevalence of hypogonadotropic hypogonadism [9, 23, 52, 53]. In addition, the increased conversion of T to estradiol by fat observed in obese men can reduce LH secretion, inducing a negative feedback on both the hypothalamus and pituitary. Accordingly, Loves et al. [54] previously reported that a weekly low dose of the aromatase inhibitor, letrozole, restored testosterone levels and increased LH levels in severely obese hypogonadal men.

Other evidence suggests that the testis could be less sensitive to gonadotropin stimulation in MetS. Pitteloud et al. [55, 56] demonstrated that human chorionic gonadotropin (hCG)-stimulated T levels were positively correlated with insulin sensitivity. Insulin itself is capable of stimulating T production while simultaneously, inhibiting SHBG concentration [57]. Hence, insulin resistance associated with obesity may contribute to the low T levels seen in obese men, due to a direct effect on the testis [55–57]. In addition, leptin exerts suppressive effects on testicular steroidogenesis and may contribute to further disruption of the pulse amplitude of GnRH [58, 59]. Finally, obesity is also associated with increased circulating TNFα levels, as a consequence of inflammatory cascade activation. Morales et al. [60] demonstrated that intratesticular TNFα delivery is associated with a blunted T response to hCG stimulation. TNFα can also inhibit LH release acting at both the pituitary and hypothalamic levels [61].

Epidemiological Association Between Testosterone and Cardiovascular Morbidity and Mortality

Despite the strong association between low T and well-known CV risk factors such as obesity, T2DM, and MetS, the relationship between CVD and serum T levels in men remains somewhat contradictory [7, 62]. Several studies have reported an association between androgen deprivation therapy for advanced prostate cancer and forthcoming CV diseases [63–66], suggesting a direct pathogenetic role of low serum testosterone. However, other reports did not confirm this association [67–70]. In addition, in a large series ($n = 1{,}687$) of subjects seeking medical care for sexual dysfunction, evaluated for up to 10 years, there was no association between incident major adverse cardiovascular diseases (MACE) and baseline T levels [71]. In the same study, however, low T, even in the overt hypogonadal range (below 8 nmol/L), was associated with an increased risk of MACE mortality [71]. In line with this data, low T has been associated with an increased mortality in patients affected by specific diseases, such as hypopituitarism [72], Klinefelter's syndrome [73], mental retardation [74], and in specific populations, such as veterans [75], and Japanese men with at least one CV factor [76]. Studies performed in community-dwelling males, however, have provided conflicting results [77–94] (see also Table 5.1). A recent meta-analysis of the available population-based studies has shown that low endogenous T levels are associated with increased risk of all-cause and CVD death,

Table 5.1 Descriptive characteristics of the available community-dwelling studies evaluating the impact of testosterone levels on forthcoming cardiovascular morbidity and mortality

Study	Name	Age (years) mean ± SD (range)	Number of subjects	Follow-up (years)	Diagnosis of hypogonadism	CV morbidity	CV mortality
Barrett-Connor and Khaw [77]	Rancho-Bernardo	40–79	1,009	12.0	TT as continuum	↔ MACE	↔ MACE
Contoreggi et al. [78]	Baltimore longitudinal study of aging	58.8 ± 10.0 34–87	170	9.5	TT as continuum	↔ IHD	NA
Yarnell et al. [79]	Caerphilly study	45–59	2,512	5.0	Lowest quintile vs. higher quintiles[a]	↔ IHD	NA
Smith et al. [80]	Caerphilly study	45–59	2,512	16.5	Lowest quintile (TT < 16.1 nmol/L) vs. higher quintiles	↔ IHD	↔ IHD
Ärnlöv et al. [81]	Framingham study	56.0 ± 12.0 30–60	2,084	10.0	Lowest quartile (TT < 11.5 nmol/L) vs. higher quartile	↔ MACE	NA
Abbott et al. [82]	Honolulu-Asia Aging study	77.6 ± 4.6 71–93	2,197	6.0	Lowest quintile (TT < 12 nmol/L) vs. highest quintiles	↔ stroke	NA
Araujo et al. [83]	Massachusetts Male Aging study	40–70	1,686	15.3	Lowest quintile (TT < 12.8 nmol/L) vs. highest quintiles	NA	↔ MACE
					Lowest quintile (FT < 0.280 nmol/L) vs. highest quintiles	NA	↓ IHD
Khaw et al. [84]	European Prospective Investigation into Cancer in Norfolk	40–79	11,606[a]	7.0	Lowest quartile (TT < 12.5 nmol/L) vs. highest quartiles	NA	↑ all causes of mortality ↑ MACE
Maggio et al. [85]	Aging in the Chianti Area study	65–92	410	6.0	BT < 2.43 nmol/L	NA	↔ All causes of mortality
Laughlin et al. [86]	Rancho-Bernardo	Median 73.6	794	11.8	Lowest quartile (TT < 8.4 nmol/L) vs. highest quartiles	NA	↑ All causes of mortality
		50–91			Lowest quartile (BT < 2.7 nmol/L) vs. highest quartiles	NA	↑ All causes of mortality ↑ MACE

Lehtonen [87]	The health of Elderly men in the city of Turku, Southwest Finland	71.5	187	10.0	TT as continuum	NA	↑ All causes of mortality
Szulc et al. [88]	MINOS study	65.4	782	10	Lowest quartile[a] vs. highest quartiles	NA	↔ All causes of mortality
Tivesten et al. [89]	Swedish Osteoporotic Fractures in Men study	75.0	3,014	4.5	Lowest quartile (TT < 11.5 nmol/L) vs. highest quartiles	NA	↑ All causes of mortality
		69–80			Lowest quartile (FT < 0.14 nmol/L) vs. highest quartiles	NA	↑ All causes of mortality
Vikan et al. [90]	Tromsø study	59.6 ± 10.2	1,568	9.1 (for morbidity observation)	Lowest quartile (TT < 9.7 nmol/L) vs. highest quartiles	↔IHD	↔ All causes of mortality ↔ MACE ↔ IHD
				11.2 (for mortality observation)	Lowest quartile (FT < 0.158 nmol/L) vs. highest quartiles	↔IHD	↑ All causes of mortality ↔ MACE ↔ IHD
Yeap et al. [91]	Health in Men study	≥74	3,443	3.5	Lowest quartile (TT < 11.7 nmol/L) vs. highest quartiles	↑ Stroke or TIA	↑ Stroke or TIA
					Lowest quartile (FT < 0.222 nmol/L) vs. highest quartile	↑ Stroke or TIA	↑ Stroke or TIA
Haring et al. [92]	Health in Men study	45–72	1,954	7.2	TT < 8.7 nmol/L	NA	↑ All causes of mortality ↑ MACE

(continued)

Table 5.1 (continued)

Study	Name	Age (years) mean ± SD (range)	Number of subjects	Follow-up (years)	Diagnosis of hypogonadism	CV morbidity	CV mortality
Menke et al. [93]	Third National Health and Nutrition Examination Survey Mortality study	≥20	1,114	9	TT ≤ difference between 90th and 10th percentiles (27;11 nmol/L)	NA	↔ All causes of mortality ↔ MACE
					FT ≤ difference between 90th and 10th percentiles (0.549;0.206 nmol/L)	NA	↑ All causes of mortality ↑ MACE
Hyde et al. [94]	Health in Men study	≥64	3,637	Mean 5.1	TT as continuum	NA	↔ All causes of mortality ↔ CV mortality
					FT as continuum	NA	↑ All causes of mortality ↑ CV mortality

TT total testosterone; *FT* free testosterone; *BT* bioavailable testosterone; *MACE* major cardiovascular events; *IHD* ischemic heart diseases; *TIA* transient ischemic attack; *NA* not available

[a]The testosterone values were not reported in the original study

but no association was observed between reduced T and incidence of CVD [62]. In addition, the same authors emphasized that there was a considerable between-study heterogeneity, which was related to study and subject characteristics, suggests that effects were driven by differences between cohorts (e.g., in underlying health status). Accordingly, meta-analysis of the available cross-sectional studies demonstrate that subjects with CVD have, on average, lower T levels than healthy controls [47]. Furthermore, several associated morbidities, such as diabetes, obesity, and hypertension, were associated with increased T differences between cases and controls, confirming numerous clinical observations. Taken together, these results suggest that low T may be considered as a marker of poor general health status, negatively affecting prognosis, rather than as a specific CV risk factor [47].

The main issue regarding the association between male hypogonadism and CVD is that it is not apparent which is the cause and which is the consequence. In fact, in prostate cancer patients, the increased prevalence of obesity, T2DM, MetS, and CVD, induced by androgen ablation, suggests that suppressed T might be the cause; however, the suppressing effect on T level of various chronic diseases (including MetS and T2DM) suggests the opposite. In line with the latter hypothesis we recently found that body mass index (BMI) played a crucial role in the stratification of hypogonadism-associated CV risk. In a retrospective analysis of a large series of subjects with sexual dysfunction, we showed that SHBG-bound and unbound T levels are inversely related to CV risk [95]. Interestingly, when obese subjects were compared to the rest of the sample, the association between T and CV risk was confirmed only in overweight and normal weight subjects. Similarly, we observed a positive association between forthcoming MACE and reduced T levels in normal weight and overweight subjects, which was lost when obese (BMI>30 kg/m^2) patients were considered [96]. Hence, while low T could contribute to the pathogenesis of CV diseases, the opposite is also conceivable. It can be speculated that obesity-associated hypogonadism is an adaptive mechanism. In fact, we cannot exclude the possibility that low T, as observed in obesity or other chronic diseases, has a protective role by turning off T-dependent functions (such as reproduction and physical vigor), that are not desirable when the organism is ill. Similar adaptive mechanisms have been previously described for other hormonal axes. A typical pattern of altered thyroid hormone metabolism, characterized by low three-iodotironin (T3) circulating levels (low T3 syndrome), has been reported in patients with different chronic illnesses, including CVD [97]. According to this view, a recent randomized controlled trial (RCT) concerning the effect of high-dose transdermal T (100 mg of a 1% gel) on more than 200 hypogonadal (total T below 12 nmol/L) frail elderly men reported a high rate of T-associated CV adverse events, which induced a premature termination of the study [98].

Biological Effects of Androgens on the Cardiovascular System

The mechanisms underlying the epidemiological associations of low T with CVD are complex and not fully understood. Both clinical and animal model evidence shows that T exerts a favorable effect upon cytokine production, vascular reactivity, inflammation, adhesion molecule expression, as well as on serum lipid concentration and haemostatic factors [7]. Atherosclerosis is a multifaceted, progressive disease affecting mainly large- and medium-sized arteries involving many components of the endocrine, vascular, metabolic, and immune systems. The complexity of the pathogenesis of atherosclerosis, and the wide range of biological actions of androgens, suggest the possibility of multiple mechanisms linking the two conditions.

In hypogonadal men with coronary artery diseases, T replacement therapy (TRT) has been reported to suppress atherogenic IL-1β, TNF-α and to increase the anti-atherogenic cytokine IL-10 [99]. Testosterone, at male physiological circulating concentrations (high nanomolar range), exerts a direct vasodilatory effect on numerous vascular beds, including the coronary arteries [100, 101]. The effects of T on the coronary arteries are thought to be mediated by the opening of large-conductance, calcium-activated, potassium channels [100, 101]. Endothelial progenitor cells play a central role in repairing and maintaining the vascular bed and other angiogenic events. T is likely to be involved in endothelial repair, since low T is associated with fewer endothelial progenitor cells in young hypogonadal men [102], whereas TRT was able to normalize their count [103]. Testosterone at physiological concentrations also has a beneficial influence on the haemostatic system through enhancement of anticoagulant activity [104].

Effects of Testosterone Replacement Therapy on CV Risk Factors and CVD

In regard to the effect of TRT on *metabolic parameters,* Isidori et al. [105], in a meta-analysis of RCTs, reported that TRT in middle-aged men reduced fat mass and total cholesterol. Similarly, Whitsel et al. [106], in a meta-analysis of the effects of intramuscular TRT in hypogonadal men, showed a small dose-dependent decrease in total cholesterol and low-density lipoprotein and HDL-cholesterol. Few RCTs have evaluated the impact of TRT in patients with MetS and T2DM. Meta-analysis of available evidence revealed that in patients with MetS, TRT was associated with a significant reduction in fasting plasma glucose, Homeostasis Model Assessment index, triglycerides and waist circumference, and an increase of HDL-cholesterol [42]. Similar results were observed when T2DM was considered. Specifically, TRT was associated with a significant reduction of fasting plasma glucose, HbA1c, fat mass, and triglycerides [16].

Concerning the role of TRT on CV events, a recent RCT study using high-dose transdermal T (100 mg of a 1% gel) on more than 200 hypogonadal (total T below 12 nmol/L) frail, elderly men showed a high rate of T-associated CV adverse events [98]. These data raised concern to warrant termination of this study. These same authors recognized that the data about the safety of TRT is limited by several factors, including that CV events were not a planned primary or secondary outcome measures and that the actual number of the adverse events was relatively small (23 vs. 5 respectively for treatment and placebo arms [98]). In contrast to this study, three other meta-analyses found no significant difference between T and placebo groups for all CV events, nor for each type of event, except for an increase in hematocrit over 50%, which was significantly more common in the T group [107–109]. Meta-analysis of specific data from studies involving subjects with CVD demonstrated that TRT was effective in men with chronic stable angina, as subjects had higher angina-free exercise tolerance than placebo-treated controls. The beneficial effects of chronic TRT on CV risk need to be further elucidated through large-scale, long-term, placebo-controlled studies [45]. Accordingly, it should be recognized that the statistical power of available meta-analyses is far from adequate, due to limited number of studies, small sample sizes, and short duration of observations.

Conclusions

Despite epidemiological evidence indicating an inverse relationship between circulating T and well-known CV risk factors (obesity, T2DM, and MetS), the relationship between CVD and serum T levels remains somewhat contradictory. It is conceivable that low T, as observed in men with chronic disease, may represent an adaptive mechanism.

References

1. National Center for Health Statistics. Health, United States, 2006: with chartbook on trends in the health of Americans. Hyattsville: National Center for Health Statistics (US); 2006.
2. Johnson FW, Gruenewald PJ, Treno AJ, et al. Drinking over the life course within gender and ethnic groups: a hyperparametric approach. J Stud Alcohol. 1998;59:568–81.
3. York JL, Welte J, Hirsch J. Gender composition of alcohol exposure on drinking occasions. J Stud Alcohol. 2003;64:790–801.
4. Miniño AM, Heron MP, Murphy SL, Kochanek KD. Centers for Disease Control and Prevention National Center for Health Statistics National Vital Statistics System. Deaths: final data for 2004. Natl Vital Stat Rep. 2007 Aug 21;55(19):1–119.
5. Barbeau EM, Krieger N, Soobader MJ. Working class matters: socioeconomic disadvantage, race/ethnicity, gender, and smoking in NHIS 2000. Am J Public Health. 2004;94:269–78.
6. Wallace Jr JM, Bachman JG, O'Malley PM, et al. Gender and ethnic differences in smoking, drinking and illicit drug use among American 8th, 10th, and 12th grade students, 1976–2000. Addiction. 2003;98:225–34.
7. Corona G, Rastrelli G, Vignozzi L, et al. Testosterone, cardiovascular disease and the metabolic syndrome. Best Pract Res Clin Endocrinol Metab. 2011;25:337–53.
8. Wu FC, von Eckardstein A. Androgens and coronary artery disease. Endocr Rev. 2003;24: 183–217.
9. Corona G, Mannucci E, Forti G, et al. Hypogonadism, ED, metabolic syndrome and obesity: a pathological link supporting cardiovascular diseases. Int J Androl. 2009;32:587–98.
10. Corona G, Mannucci E, Fisher AD, et al. Low levels of androgens in men with erectile dysfunction and obesity. J Sex Med. 2008;5:2454–63.
11. Saad F, Aversa A, Isidori AM, et al. Testosterone as potential effective therapy in treatment of obesity in men with testosterone deficiency: a review. Curr Diabetes Rev. 2012;23:8.

12. Traish AM, Guay A, Feeley R, et al. The dark side of testosterone deficiency: I. Metabolic syndrome and erectile dysfunction. J Androl. 2009;30:10–22.
13. Giagulli VA, Kaufman JM, Vermeulen A. Pathogenesis of the decreased androgen levels in obese men. J Clin Endocrinol Metab. 1994;79:997–1000.
14. Vermeulen A, Kaufman JM, Deslypere JP, et al. Attenuated luteinizing hormone (LH) pulse amplitude but normal LH pulse frequency, and its relation to plasma androgens in hypogonadism of obese men. J Clin Endocrinol Metab. 1993;76:1140–6.
15. Traish AM, Saad F, Guay A. The dark side of testosterone deficiency: II. Type 2 diabetes and insulin resistance. J Androl. 2009;30:23–32.
16. Corona G, Monami M, Rastrelli G, et al. Type 2 diabetes mellitus and testosterone: a meta-analysis study. Int J Androl. 2011;34:528–40.
17. Wang C, Jackson G, Jones TH, et al. Low testosterone associated with obesity and the metabolic syndrome contributes to sexual dysfunction and cardiovascular disease risk in men with type 2 diabetes. Diabetes Care. 2011;34:1669–75.
18. Ding EL, Song Y, Malik VS, et al. Sex differences of endogenous sex hormones and risk of type 2 diabetes: a systematic review and meta-analysis. JAMA. 2006;295:1288–99.
19. Corona G, Mannucci E, Forti G, et al. Following the common association between testosterone deficiency and diabetes mellitus, can testosterone be regarded as a new therapy for diabetes? Int J Androl. 2009;32:431–41.
20. Anderson SG, Heald A, Younger N, et al. Screening for hypogonadism in diabetes 2008/9: results from the Cheshire Primary Care cohort. Prim Care Diabetes. 2012;6:143–8.
21. Wang C, Nieschlag E, Swerdloff R, et al. Investigation, treatment and monitoring of late-onset hypogonadism in males: ISA, ISSAM, EAU, EAA and ASA recommendations. Eur J Endocrinol. 2008;159:507–14.
22. Buvat J, Maggi M, Gooren L, et al. Endocrine aspects of male sexual dysfunctions. In: Montorsi F, Basson R, Adaikan G, Becher E, Clayton A, Giuliano G, Khoury S, Sharlip I, editors. Sexual medicine, sexual dysfunctions in men and women. Proceedings of the 3rd international consultation on sexual medicine. Paris: Health Publication Ltd; 2010. p. 681.
23. Dhindsa S, Prabhakar S, Sethi M, et al. Frequent occurrence of hypogonadotropic hypogonadism in type 2 diabetes. J Clin Endocrinol Metab. 2004;89:5462–8.
24. Kumai T, Tanaka M, Watanabe M, et al. Possible involvement of androgen in increased norepinephrine synthesis in blood vessels of spontaneously hypertensive rats. Jpn J Pharmacol. 1994;66:439–44.
25. Cambotti LJ, Cole FE, Gerall AA, et al. Neonatal gonadal hormones and blood pressure in the spontaneously hypertensive rat. Am J Physiol. 1984;247:E258–64.
26. Reckelhoff JF, Zhang H, Granger JP. Testosterone exacerbates hypertension and reduces pressure-natriuresis in male spontaneously hypertensive rats. Hypertension. 1998;31:435–9.
27. Yanes LL, Iliescu R, Sartori-Valinotti JC, et al. Testosteronedependent hypertension and upregulation of intrarenal angiotensinogen in Dahl salt-sensitive rats. Am J Physiol Renal Physiol. 2009;296:F771–9.
28. Hughes GS, Mathur RS, Margolius HS. Sex steroid hormones are altered in essential hypertension. J Hypertens. 1989;7:181–7.
29. Svartberg J, von Mühlen D, Schirmer H, et al. Association of endogenous testosterone with blood pressure and left ventricular mass in men. The Tromsø Study. Eur J Endocrinol. 2004;150:65–71.
30. Labropoulos B, Velonakis E, Oekonomakos P, et al. Serum sex hormones in patients with coronary disease and their relationship to known factors causing atherosclerosis. Cardiology. 1982;69:98–103.
31. Corona G, Rastrelli G, Morelli A, et al. Hypogonadism and metabolic syndrome. J Endocrinol Invest. 2011;34:557–67.
32. Corona G, Mannucci E, Lotti F, et al. Pulse pressure, an index of arterial stiffness, is associated with androgen deficiency and impaired penile blood flow in men with ED. J Sex Med. 2009;6:285–93.

33. Fogari R, Preti P, Zoppi A, et al. Serum testosterone levels and arterial blood pressure in the elderly. Hypertens Res. 2005;28:625–30.
34. Haffner SM, Mykkanen L, Valdez RA, et al. Relationship of sex hormones to lipids and lipoproteins in nondiabetic men. The Journal of clinical endocrinology and metabolism. 1993;77:1610–5.
35. Malkin CJ, Pugh PJ, Jones RD, et al. The effect of testosterone replacement on endogenous inflammatory cytokines and lipid profiles in hypogonadal men. The Journal of clinical endocrinology and metabolism. 2004;89:3313–8.
36. Braga-Basaria M, Muller DC, Carducci MA, et al. Lipoprotein profile in men with prostate cancer undergoing androgen deprivation therapy. International journal of impotence research. 2006;18:494–8.
37. Smith MR, Finkelstein JS, McGovern FJ, et al. Changes in body composition during androgen deprivation therapy for prostate cancer. The Journal of clinical endocrinology and metabolism. 2002;87:599–603.
38. Yannucci J, Manola J, Garnick MB, et al. The effect of androgen deprivation therapy on fasting serum lipid and glucose parameters. The Journal of urology. 2006;176:520–5.
39. Dockery F, Bulpitt CJ, Agarwal S, et al. Testosterone suppression in men with prostate cancer leads to an increase in arterial stiffness and hyperinsulinaemia. Clin Sci (Lond). 2003;104:195–201.
40. Moorjani S, Dupont A, Labrie F, et al. Changes in plasma lipoproteins during various androgen suppression therapies in men with prostatic carcinoma: effects of orchiectomy, estrogen, and combination treatment with luteinizing hormone-releasing hormone agonist and flutamide. The Journal of clinical endocrinology and metabolism. 1988;66:314–22.
41. Smith JC, Bennett S, Evans LM, et al. The effects of induced hypogonadism on arterial stiffness, body composition, and metabolic parameters in males with prostate cancer. The Journal of clinical endocrinology and metabolism. 2001;86:4261–7.
42. Corona G, Monami M, Rastrelli G, et al. Testosterone and metabolic syndrome: a meta-analysis study. J Sex Med. 2011;8:272–83.
43. Brand JS, van der Tweel I, Grobbee DE, et al. Testosterone, sex hormone-binding globulin and the metabolic syndrome: a systematic review and meta-analysis of observational studies. Int J Epidemiol. 2011;40:189–207.
44. Wu FC, Tajar A, Beynon JM, et al. Identification of late-onset hypogonadism in middle-aged and elderly men. N Engl J Med. 2010;363:123–35.
45. Singh R, Artaza JN, Taylor WE, et al. Testosterone inhibits adipogenic differentiation in 3T3-L1 cells: nuclear translocation of androgen receptor complex with beta-catenin and T-cell factor 4 may bypass canonical Wnt signaling to down-regulate adipogenic transcription factors. Endocrinology. 2006;147:141–54.
46. Corona G, Rastrelli G, Monami M, et al. Hypogonadism as a risk factor for cardiovascular mortality in men: a meta-analytic study. Eur J Endocrinol. 2011;165:687–701.
47. Marin P, Oden B, Bjorntorp P. Assimilation and mobilization of triglycerides in subcutaneous abdominal and femoral adipose tissue in vivo in men: effects of androgens. J Clin Endocrinol Metab. 1995;80:239–43.
48. Corona G, Forti G, Maggi M. Why can patients with erectile dysfunction be considered lucky? The association with testosterone deficiency and metabolic syndrome. Aging Male. 2008;11:193–9.
49. Zhang XH, Filippi S, Morelli A, et al. Testosterone restores diabetes-induced erectile dysfunction and sildenafil responsiveness in two distinct animal models of chemical diabetes. J Sex Med. 2006;3:253–64.
50. Vignozzi L, Morelli A, Filippi S, et al. Testosterone regulates RhoA/Rho-kinase signaling in two distinct animal models of chemical diabetes. J Sex Med. 2007;4:620–30.
51. Brüning JC, Gautam D, Burks DJ, et al. Role of brain insulin receptor in control of body weight and reproduction. Science. 2007;289:2122–5.
52. Corona G, Mannucci E, Petrone L, et al. Association of hypogonadism and type II diabetes in men attending an outpatient erectile dysfunction clinic. Int J Impot Res. 2006;18:190–7.

53. Corona G, Mannucci E, Schulman C, et al. Psychobiologic correlates of the metabolic syndrome and associated sexual dysfunction. Eur Urol. 2006;50:595–604.
54. Loves S, Ruinemans-Koerts J, de Boer H. Letrozole once a week normalizes serum testosterone in obesity-related male hypogonadism. Eur J Endocrinol. 2008;158:741–7.
55. Pitteloud N, Hardin M, Dwyer AA, et al. Increasing insulin resistance is associated with a decrease in Leydig cell testosterone secretion in men. J Clin Endocrinol Metab. 2005;90: 2636–41.
56. Pitteloud N, Dwyer AA, DeCruz S, et al. Inhibition of LH secretion by testosterone in men requires aromatization for its pituitary but not its hypothalamic effects: evidence from the tandem study of normal and gonadotrophin-releasing hormone-deficient men. J Clin Endocrinol Metab. 2008;93:784–91.
57. Pasquali R, Casimirri F, De Iasio R, et al. Insulin regulates testosterone and sex hormone-binding globulin concentrations in adult normal weight and obese men. J Clin Endocrinol Metab. 1995;80:654–8.
58. Isidori AM, Caprio M, Strollo F, et al. Leptin and androgens in male obesity: evidence for leptin contribution to reduced androgen levels. J Clin Endocrinol Metab. 1999;84:3673–80.
59. Isidori AM, Strollo F, Morè M, et al. Leptin and aging: correlation with endocrine changes in male and female healthy adult populations of different body weights. J Clin Endocrinol Metab. 2000;85:1954–62.
60. Morales V, Santana P, Díaz R, et al. Intratesticular delivery of tumor necrosis factor-alpha and ceramide directly abrogates steroidogenic acute regulatory protein expression and Leydig cell steroidogenesis in adult rats. Endocrinology. 2003;144:4763–72.
61. McEwan DJ. Interactions between TNF and GnRH. Neurochem Res. 2008;33:678–82.
62. Araujo AB, Dixon JM, Suarez EA, et al. Clinical review: endogenous testosterone and mortality in men: a systematic review and meta-analysis. J Clin Endocrinol Metab. 2011;96:3007–19.
63. Keating NL, O'Malley AJ, Smith MR. Diabetes and cardiovascular disease during androgen deprivation therapy for prostate cancer. J Clin Oncol. 2006;24:4448–56.
64. Saigal CS, Gore JL, Krupski TL, et al. Androgen deprivation therapy increases cardiovascular morbidity in men with prostate cancer. Cancer. 2007;110:1493–500.
65. D'Amico AV, Denham JW, Crook J, et al. Influence of androgen suppression therapy for prostate cancer on the frequency and timing of fatal myocardial infarctions. J Clin Oncol. 2007;25:2420–5.
66. Tsai HK, D'Amico AV, Sadetsky N, et al. Androgen deprivation therapy for localized prostate cancer and the risk of cardiovascular mortality. J Natl Cancer Inst. 2007;99:1516–24.
67. Efstathiou JA, Bae K, Shipley WU, et al. Cardiovascular mortality and duration of androgen deprivation for locally advanced prostate cancer: analysis of RTOG 92–02. Eur Urol. 2008;54:816–23.
68. Efstathiou JA, Bae K, Shipley WU, et al. Cardiovascular mortality after androgen deprivation therapy for locally advanced prostate cancer: RTOG 85–31. J Clin Oncol. 2009;27:92–9.
69. Roach 3rd M, Bae K, Speight J, et al. Short-term neoadjuvant androgen deprivation therapy and external-beam radiotherapy for locally advanced prostate cancer: long-term results of RTOG 8610. J Clin Oncol. 2008;26:585–91.
70. Bolla M, de Reijke TM, Van Tienhoven G, et al. Duration of androgen suppression in the treatment of prostate cancer. N Engl J Med. 2009;360:2516–27.
71. Corona G, Monami M, Boddi V, et al. Low testosterone is associated with an increased risk of MACE lethality in subjects with erectile dysfunction. J Sex Med. 2010;7:1557–64.
72. Tomlinson JW, Holden N, Hills RK, et al. Association between premature mortality and hypopituitarism. West Midlands Prospective Hypopituitary Study Group. Lancet. 2001;357:425–31.
73. Bojesen A, Juul S, Birkebaek N, et al. Increased mortality in Klinefelter syndrome. J Clin Endocrinol Metab. 2004;89:3830–4.
74. Hamilton JB, Mestler GE. Mortality and survival: comparison of eunuchs with intact men and women in a mentally retarded population. J Gerontol. 1969;24:395–411.

75. Shores MM, Matsumoto AM, Sloan KL, et al. Low serum testosterone and mortality in male veterans. Arch Intern Med. 2006;166:1660–5.
76. Akishita M, Hashimoto M, Ohike Y, et al. Low testosterone level as a predictor of cardiovascular events in Japanese men with coronary risk factors. Atherosclerosis. 2010;210:232–6.
77. Barrett-Connor E, Khaw KT. Endogenous sex hormones and cardiovascular disease in men. A prospective population-based study. Circulation. 1988;78:539–45.
78. Contoreggi CS, Blackman MR, Andres R, et al. Plasma levels of estradiol, testosterone, and DHEAS do not predict risk of coronary artery disease in men. J Androl. 1990;11:460–70.
79. Yarnell JW, Beswick AD, Sweetnam PM, et al. Endogenous sex hormones and ischemic heart disease in men. The Caerphilly prospective study. Arterioscler Thromb Vasc Biol. 1993;13:517–20.
80. Smith GD, Ben-Shlomo Y, Beswick A, et al. Cortisol, testosterone, and coronary heart disease: prospective evidence from the Caerphilly Study. Circulation. 2005;112:332–40.
81. Ärnlöv J, Pencina MJ, Amin S, et al. Endogenous sex hormones and cardiovascular disease incidence in men. Ann Intern Med. 2006;145:176–84.
82. Abbott RD, Launer BL, Rodriguez GW, et al. Serum estradiol and risk of stroke in elderly men. Neurology. 2007;68:563–8.
83. Araujo AB, Kupelian V, Page ST, et al. Sex steroids and all-cause and cause-specific mortality in men. Arch Intern Med. 2007;167:1252–60.
84. Khaw KT, Dowsett M, Folkerd E, et al. Endogenous testosterone and mortality due to all causes, cardiovascular disease, and cancer in men: European prospective investigation into cancer in Norfolk (EPIC-Norfolk) prospective population study. Circulation. 2007;116:2694–701.
85. Maggio M, Lauretani F, Ceda GP, et al. Relationship between low levels of anabolic hormones and 6-year mortality in older men: the Aging in the Chianti Area (INCHIANTI) study. Arch Intern Med. 2007;167:2249–54.
86. Laughlin GA, Barrett-Connor E, Bergstrom J. Low serum testosterone and mortality in older men. J Clin Endocrinol Metab. 2008;93:68–75.
87. Lehtonen A, Huupponen R, Tuomilehto J, et al. Serum testosterone but not leptin predicts mortality in elderly men. Age Ageing. 2008;37:461–4.
88. Szulc P, Claustrat B, Delmas PD. Serum concentrations of 17β-E2 and 25 hydroxycholecalciferol (25OHD) in relation to all cause mortality in older men—the MINOS study. Clin Endocrinol (Oxf). 2009;71:594–602.
89. Tivesten A, Vandenput L, Labrie F, et al. Low serum testosterone and estradiol predict mortality in elderly men. J Clin Endocrinol Metab. 2009;94:2482–8.
90. Vikan T, Schirmer H, Njølstad I, et al. Endogenous sex hormones and the prospective association with cardiovascular disease and mortality in men: the Tromsø Study. Eur J Endocrinol. 2009;161:435–42.
91. Yeap BB, Hyde Z, Almeida OP, et al. Lower testosterone levels predict incident stroke and transient ischemic attack in older men. Clin Endocrinol Metab. 2009;94:2353–9.
92. Haring R, Völzke H, Steveling A, et al. Low serum testosterone levels are associated with increased risk of mortality in a population-based cohort of men aged 20–79. Eur Heart J. 2010;31:1494–501.
93. Menke A, Guallar E, Rohrmann S, et al. Sex steroid hormone concentrations and risk of death in US men. Am J Epidemiol. 2010;171:583–92.
94. Hyde Z, Norman PE, Flicker L, et al. Low free testosterone predicts mortality from cardiovascular disease but not other causes: the health in men study. J Clin Endocrinol Metab. 2012;97:179–89.
95. Corona G, Rastrelli G, Balercia G, Sforza A, Forti G, Maggi M.Testosterone and cardiovascular risk in patients with erectile dysfunction. J Endocrinol Invest. 2011 Nov 8. [Epub ahead of print].
96. Corona G, Rastrelli G, Monami M, et al. Body mass index regulates hypogonadism-associated CV risk: results from a cohort of subjects with erectile dysfunction. J Sex Med. 2011;8:2098–105.

97. Iervasi G, Pingitore A, Landi P, et al. Low-T3 syndrome: a strong prognostic predictor of death in patients with heart disease. Circulation. 2003;107:708–13.
98. Basaria S, Coviello AD, Travison TG, et al. Adverse events associated with testosterone administration. N Engl J Med. 2010;363:109–22.
99. Jones TH. Testosterone deficiency: a risk factor for cardiovascular disease? Trends Endocrinol Metab. 2010;21:496–503.
100. Malkin CJ, Pugh PJ, Morris PD, et al. Testosterone replacement in hypogonadal men with angina improves ischaemic threshold and quality of life. Heart. 2004;90:871–6.
101. webb CM, Adamson DL, de Zeigler D, et al. Effect of acute testosterone on myocardial ischemia in men with coronary artery disease. Am J Cardiol. 1999;83:437–9.
102. Foresta C, Caretta N, Lana A, et al. Reduced number of circulating endothelial progenitor cells in hypogonadal men. J Clin Endocrinol Metab. 2006;91:4599–602.
103. Foresta C, Di Mambro A, Caretta N, et al. Effect of vardenafil on endothelial progenitor cells in hypogonadotrophic hypogonadal patients: role of testosterone treatment. Clin Endocrinol. 2008;71:412–6.
104. Jin H, Lin J, Fu L, et al. Physiological testosterone stimulates tissue plasminogen activator and tissue factor pathway inhibitor and inhibits plasminogen activator inhibitor type 1 release in endothelial cells. Biochem Cell Biol. 2007;85:246–51.
105. Isidori AM, Giannetta E, Greco EA, et al. Effects of testosterone on body composition, bone metabolism and serum lipid profile in middle-aged men: a meta-analysis. Clin Endocrinol (Oxf). 2005;63:280–93.
106. Whitsel EA, Boyko EJ, Matsumoto AM, et al. Intramuscular testosterone esters and plasma lipids in hypogonadal men: a meta-analysis. Am J Med. 2001;111:261–9.
107. Haddad RM, Kennedy CC, Caples SM, et al. Testosterone and cardiovascular risk in men: a systematic review and meta-analysis of randomized placebo-controlled trials. Mayo Clin Proc. 2007;82:29–39.
108. Fernández-Balsells MM, Murad MH, Lane M, et al. Clinical review 1: adverse effects of testosterone therapy in adult men: a systematic review and meta-analysis. J Clin Endocrinol Metab. 2010;95:2560–75.
109. Calof OM, Singh AB, Lee ML, et al. Adverse events associated with testosterone replacement in middle-aged and older men: a meta-analysis of randomized, placebo-controlled trials. J Gerontol A Biol Sci Med Sci. 2005;60:1451–7.

Chapter 6
The Role of Androgens in Prostate Cancer

Frances Alba, Claudio A. Romero, and Run Wang

Background

Prostate cancer is the most common non-skin cancer in American males and the second most common cause of cancer-related death in men. With the advent of prostate-specific antigen (PSA) screening in the late 1980s, an initial decrease in the incidence of the disease was observed. However, recent evidence suggests that there is now a trend towards increasing incidence, while death from prostate cancer continues to decline [1].

The prostate is an androgen-responsive gland that surrounds the urethra just below the bladder. At a cellular level, the prostate is composed of epithelial and stromal elements. Both of these cells types express the androgen receptor (AR) and are the primary functional units of the prostate. Among other various proteins and enzymes that are involved in cell cycle regulation and cell survival, the epithelial cells secrete PSA and prostate-specific phosphatase, which constitute the major enzymatic portion of seminal fluid [2]. A very complex relationship exists between stromal and epithelial cells, and they communicate through hormonal, paracrine, and autocrine activity. Under the influence of androgen stimulation, they regulate cell growth and glandular development [3, 4]. The tightly controlled interaction between the prostate cell types maintains homeostasis, such that only 0.2% of normal prostatic epithelial cells proliferate and undergo apoptosis per day [5]. In fact, because of this

F. Alba, MD • C.A. Romero, MD
Department of Urology, University of Texas Health Science Center, Houston, TX, USA
e-mail: frances.m.alba@uth.tmc.edu

R. Wang, MD, FACS (✉)
Department of Urology, University of Texas Medical School at Houston and MD Anderson Cancer Center, 6431 Fannin Street, MSB 6.018, Houston, TX 77030, USA
e-mail: run.wang@uth.tmc.edu

W.J.G. Hellstrom (ed.), *Androgen Deficiency and Testosterone Replacement: Current Controversies and Strategies*, Current Clinical Urology,
DOI 10.1007/978-1-62703-179-0_6, © Springer Science+Business Media, LLC 2013

tight balance between proliferation and cell death in the normal prostate tissue, the volume of the gland is maintained, even in the setting of additional androgenic stimulation such as with testosterone replacement [6].

Synthesis and Metabolism of Androgens

Androgens are male sex hormones that control the embryological development of the prostate, growth and function of male reproductive organs, and development of male secondary characteristics. Conversely, the absence of androgens, by castration, for example, causes diffuse atrophy primarily of the luminal epithelial cells, but not the stromal cells, resulting in involution of the prostate gland [6].

Testosterone and 5-alpha-dihydrotestosterone (DHT) are the two most essential androgens in men. Testosterone is the major androgen in circulation, whereas DHT is the predominate androgen within prostate tissue [2, 7]. Testosterone synthesis is tightly controlled by the hypothalamic–pituitary–gonadal axis (Fig. 6.1). The hypothalamus releases luteinizing hormone-releasing hormone (LHRH). In response to LHRH, the anterior pituitary releases luteinizing hormone (LH). The LH is transported to the testes where it acts directly on the Leydig cells and stimulates testosterone synthesis. Through negative feedback inhibition, testosterone acts on the hypothalamus and pituitary glands to decrease secretion of LHRH and LH [4].

The Leydig cells of the testes produce and release about 90% of the circulating testosterone [8, 9]. The remaining 5–10% comes from the adrenal cortex under stimulation by adrenocorticotropic hormone from the pituitary gland. Under normal physiologic conditions, androgens produced by the adrenal gland do not significantly influence prostate growth. Once in circulation, testosterone is primarily protein-bound, either loosely to albumin or strongly to sex hormone binding globulin (SHBG). The concentration of SHBG determines the bioavailability of circulating free testosterone to tissues. It binds to testosterone in the blood, thereby reducing the extent to which testosterone is free to cross cell membranes and enter target cells. Unbound, or free testosterone comprises only about 2% of the total serum testosterone and is the most biologically active form of testosterone [7, 10].

Free testosterone is lipophilic and can easily diffuse across cell membranes into its target cells within the prostate. Upon entry into prostate tissues, it can bind directly to the AR and influence prostate cell differentiation. Alternatively, it can be rapidly and irreversibly converted by the enzyme 5-alpha-reductase to DHT. DHT is a much more potent androgen in promoting prostate growth, because it is produced at low circulating levels of testosterone [11, 12]. Thus, a major role of 5-alpha-reductase may be to ensure normal prostate function even at low testosterone levels [14].

There are two known types of the enzyme 5-alpha-reductase. Type 1 primarily exists in the skin, liver, and, to a lesser degree, in prostate. Type 2 predominates in the prostate stroma and epithelium and is important for sexual and prostate development early in life and to prostate hyperplasia in adulthood [4]. Individuals with a genetic mutation causing deficiency in 5-alpha-reductase have high concentrations of circulating testosterone and decreased DHT. They may express ambiguous

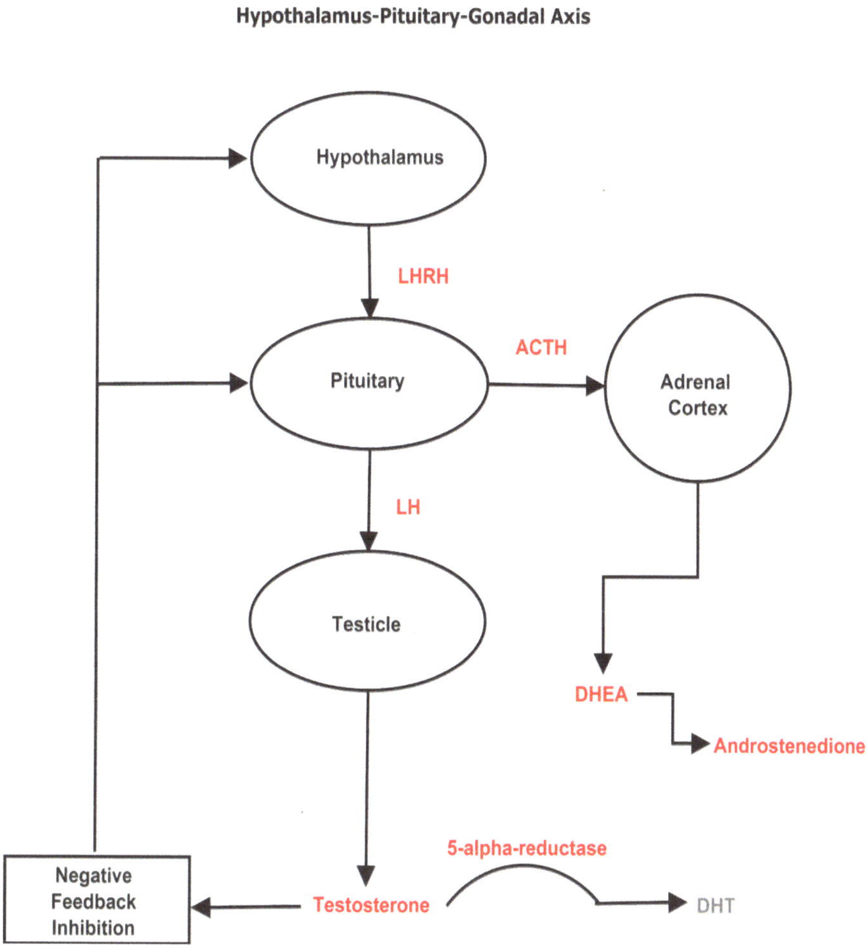

Fig. 6.1 Simplified endocrinology of the prostate. Luteinizing hormone-releasing hormone (LHRH) from the hypothalamus stimulates the anterior pituitary to secrete luteinizing hormone (LH), which stimulates the Leydig cells of the testicles to produce and release testosterone. Through negative feedback inhibition, testosterone acts on the hypothalamus and pituitary gland to decrease secretion of LHRH and LH. Adrenocorticotropic hormone (ACTH) from the pituitary stimulates the adrenal gland to release dehydroepiandrosterone (DHEA) and other less important androgens

genitalia at birth and virilization at puberty. However, their prostate remains small and they do not develop BPH or prostate cancer later in life [4]. This finding underscores the significance of DHT and 5-alpha-reductase in prostate development and prostate disease.

In addition, multiple studies have demonstrated that development and progression of prostate cancer is linked to increased prostatic expression of 5-alpha-reductase. Specifically, prostate cancer development is associated with a relative imbalance in the expression of type 1 vs. type 2 5-alpha-reductase compared to benign tissue [13]. This imbalance may prove to be relevant when evaluating the different 5-alpha-reductase inhibitors and their effect on the prevention and treatment of prostate disease.

Finasteride and dutasteride are inhibitors of 5-alpha-reductase and have been shown to decrease levels of DHT in the plasma and in the prostate. Compared to finasteride, which only inhibits type 2, dutasteride inhibits both isoforms of 5-alpha-reductase. Treatment with these medications induces a decrease in prostate volume, PSA, and has been proven to reduce the symptoms associated with benign prostatic hypertrophy [14–16].

The application of 5-alpha-reductase inhibitors toward the treatment and prevention of prostate cancer has also been studied. Thompson et al. conducted the Prostate Cancer Prevention Trial and demonstrated that finasteride reduced the risk of prostate cancer by 25% compared with placebo, albeit with a slightly higher risk of detecting high grade (Gleason 7–10) disease of 6.4% vs. 5.1% [17]. There are other recent trials that have further demonstrated that 5-alpha-reductase inhibitors reduce the risk of prostate cancer in patients who are at high risk for the disease [18]. Clearly, 5-alpha-reductase and its role in metabolism of androgens is an important player in the pathogenesis of prostate cancer.

In addition to binding with higher affinity to the prostatic AR, DHT is present in higher concentration within prostate cell cytoplasm, comprising about 90% of the total androgen concentration within the cell. Binding of DHT to the intracytoplasmic AR induces an active translocation of the steroid ligand–receptor complex into the nucleus where it binds to DNA and activates androgen response elements [14]. This incites the production of hundreds of growth factors, cytokines and other hormones that are responsible for balancing cell turnover, influencing epithelial cell differentiation and production of prostatic secretions such as PSA and prostate-specific phosphatase [2].

In healthy prostates, androgens and AR regulate prostatic maintenance and survival, whereas in prostate cancer they function as inducers of uncontrolled cell growth. During the initial phases of the disease, malignant prostate cancer cells are usually androgen dependent. Therefore, they are subject to treatment with androgen deprivation to castrate levels, which will effectively inhibit the growth of these tumors. Eventually, the cancer will recur in a castration-resistant form and will no longer respond to additional hormonal manipulation [4]. Even though the cancer no longer responds to treatment, the AR is still of vital importance to the growth and survival of the cancer cells. However, castrate-resistant prostate cancer is beyond the scope of this chapter.

Historical Perspective

There are many available effective treatment options for patients with locally confined prostate cancer including surgical and radiation modalities. The beneficial effects of androgen deprivation on men with prostate cancer were first realized over

70 years ago, and today, it remains a mainstay of treatment for advanced prostate cancer patients [19].

In 1941, Huggins and Hodges established a strong link between prostate cancer and testosterone. Twenty-five years later, they won a Nobel Prize for their groundbreaking work in prostate cancer treatment. They observed increased levels of acid phosphatase in two patients with metastatic prostate cancer who received exogenous testosterone, although one of the patients was already castrated. They also described disease regression in patients with metastatic prostate cancer after the administration of high doses of estrogens or surgical castration. They correlated this with a decrease in acid phosphatase level. Despite the multiple methodological flaws in the study, most notably the small cohort size, they arrived at the conclusion that high levels of serum testosterone were associated with enhanced prostate cancer growth [20]. With these observations, they provided the first scientific basis for advanced prostate cancer treatment.

Androgens and Development of Prostate Cancer

Subsequent studies, during the pre-PSA era, supported the correlation between testosterone and prostate cancer progression. Prout et al. reported their results of an observational study of patients with recurrent prostate cancer after castration. The patients were administered exogenous testosterone, and this resulted in disease progression or death [21]. Similarly, Fowler et al. also reported on the adverse effects of testosterone administration to patients with metastatic prostate cancer. However, the majority of patients in this study were already androgen deprived upon receiving testosterone administration [22].

These early observations gave rise to the concept that high testosterone results in growth of prostate cancer and administration of exogenous testosterone to men with prostate cancer is harmful. Further research went beyond the correlation and sought to prove causation. For example, Pollard et al. supported this concept by inducing prostate cancer in rat models by exposing them to exogenous testosterone [23]. While there may be compelling evidence in animal and laboratory models, quality evidence in humans is lacking [24].

Although androgens are important for maintaining the prostate gland, and despite the effectiveness of androgen deprivation therapy against metastatic prostate cancer, the hypothesis of *high* circulating testosterone levels adversely affecting prostate cancer has been challenged in recent literature. Some studies have found a positive correlation between prostate cancer and testosterone levels; [25, 26] however, the vast majority has found no direct relationship between high levels of testosterone and the pathogenesis of prostate cancer [27, 28].

Roddam et al. has demonstrated the lack of evidence based support for this concept after reviewing 18 prospective studies which evaluated the association between testosterone level and prostate cancer risk. Clinical data from 3,886 men with prostate cancer and 6,438 control subjects were analyzed. They found that serum testosterone was not significantly higher in patients with prostate cancer. Conversely, the incidence of prostate cancer was not significantly increased in patients with high testosterone levels compared to those with low testosterone. Although the validity of some of the individual studies is in question due to methodological issues, this meta-analysis concluded that there was no statistical difference in pre-diagnostic serum androgen concentration between patients who developed prostate cancer and control groups. Therefore, high androgen concentration is not associated with the risk of prostate cancer [28].

Contrary to the above reports, a recently published prospective observational study by Pierorazio et al. evaluated 781 men with hormone measurements over several decades. They observed the likelihood of high risk prostate cancer in men over age 65 was double for each unit rise in the free testosterone index. The authors concluded that higher levels of serum free testosterone are associated with an increased risk of aggressive prostate cancer in older men. This study differed from others in the literature because it considered the effect of long term exposure to circulating androgens and its importance in the development of prostate cancer [29].

The opposite spectrum of the association has also been evaluated. One would expect that if high androgen levels increased the risk of prostate cancer, then it would follow that low androgen levels would decrease the risk. Surprisingly, however, multiple studies have not shown this to be true. There are numerous examples in the literature that have demonstrated low androgen levels not only do not decrease the risk, but may actually be associated with an increased risk of prostate cancer, worse 5-year biochemical relapse-free survival, increased positive surgical margins, worse pathological stage, increased percentage positive biopsy cores, and higher Gleason score [29–32].

In an early study alluding to this concept, Morgentaler et al. observed an increased rate of prostate cancer in hypogonadal men (mean age 58 years) with a normal prostate exam and a PSA < 4 ng/mL. In follow-up investigations, including a larger sample (345 men) of patients with hypogonadism and PSA < 4 ng/mL, the same authors found a correlation between the severity of testosterone deficiency and an increased risk for a positive prostate biopsy. There was more than double the risk of positive biopsy in patients with testosterone levels in the lowest tertile compared to those in the highest tertile [33].

In addition to the described correlations between low testosterone level and cancer risk, there is also evidence for the associations with worse pathological determinants of prostate cancer [34–36]. Zhang et al. measured serum total and free testosterone levels in 164 patients with high or moderate grade prostate cancer. The levels in patients with high grade prostate cancer were significantly lower than in patients with moderate grade cancer [37]. Another study, by Massengill et al., retrospectively analyzed the records from a large group of patients who had undergone radical prostatectomy for prostate cancer. They found that preoperative testosterone was significantly lower in patients with non-organ confined disease compared to those with organ confined disease. They concluded that pretreatment total testosterone level was a predictor of extraprostatic disease in patients with localized prostate cancer [38].

It has been postulated that aggressive prostate cancer can exist and may grow with a low intraprostatic DHT concentration such as that induced by 5-alpha-reductase medication [39]. As previously discussed, treatment with 5-alpha-reductase inhibitors reduces the intraprostatic concentration of DHT. The results of the Prostate Cancer Prevention Trial established a link between androgen metabolism and prostate cancer. Inhibition of 5-alpha-reductase and, by direct correlation, reduction in intraprostatic DHT levels, are associated with decreased incidence of low grade prostate cancer and slightly higher incidence of high grade cancer [17]. Bologna et al. assessed the effects of finasteride, cyproterone, and hydroxyflutamide on prostate cancer tissue culture and reported that cancer cells stimulated by testosterone and DHT show growth only at low concentrations of androgens. If higher concentrations were used, prostate cancer cell growth was inhibited [40].

Nishiyama et al. reported that patients with a Gleason score of 7–10 prostate cancer had lower concentrations of intraprostatic DHT compared to Gleason 6 cancers. However, there was no difference in serum androgens between the patient groups. They concluded that low DHT in cases of aggressive prostate cancer is probably sufficient to activate AR expression and propagate tumor growth [41].

Saturation Theory

Prostate cancer regression at castration levels of testosterone is undeniable. Cancer growth and progression after administration of testosterone to patients with prostate cancer and castration levels of testosterone is also an observable fact. However, it has never been reliably demonstrated that an increase in testosterone levels, above a near castrate level, causes any significant incremental growth in prostate cancer. To explain these observations, investigators have introduced the "saturation model," which proposes that there is a certain serum androgen concentration where all of the available AR's are bound to androgen [41]. The receptors become saturated with ligand and any additional androgen above this saturation point is unable to further stimulate prostate cancer growth. This saturation point occurs at low concentrations of androgen, slightly above the castrate range. This interaction between androgen and the AR is similar to other biological systems in which receptors become saturated by their ligands. There are two phases in the saturation model. At or below near castrate levels of testosterone, prostate cancer cells are the most sensitive to androgens such that androgen is the rate limiting factor in the AR activation pathway. Above this level, prostate cells exhibit little if any growth changes and become indifferent to higher levels of androgens [41].

A study by Marks et al. gives strong support for the saturation effect on the AR. Forty-four patients with hypogonadism were randomized to testosterone replacement or placebo for 6 months. While there was an appropriate increase in serum testosterone levels in the study group patients, no change was observed in prostatic tissue testosterone or DHT levels. In addition, there was no change in tissue biomarkers, gene expression, prostate histology, or incidence of cancer between the groups. PSA levels increased slightly in both groups, although it was higher at baseline in the group receiving testosterone [42].

Suppresion Theory

As described previously, a number of studies have established an association between high-risk prostate cancer and low serum testosterone levels. While there is a correlation, there is not enough evidence to conclude if one is cause or consequence of the other. In an attempt to explain this association, a suppression theory has been proposed. Based on measurements of preoperative and postoperative testosterone levels in men undergoing radical prostatectomy, Miller et al. hypothesized that prostate cancer cells may secrete a substance that interferes with the normal secretion of testosterone by the testis. Investigators also measured estradiol, free testosterone, DHT, FSH, and LH. There was an increase in the postoperative levels of all these hormones except for DHT, which was decreased relative to the preoperative levels. Given these findings, the authors suggested that there may be a factor, such as inhibin or DHT, secreted by prostate cancer cells causing suppression at the hypothalamic–pituitary axis, ultimately resulting in a decrease in testosterone production [43].

Recently, several studies have been published that lend support for the suppression theory [37, 44]. Yamamoto et al. measured testosterone levels before and after surgery in a cohort of 272 patients undergoing radical prostatectomy. Comparative postoperative levels were available in 222 patients. Forty-nine patients had a serum testosterone level lower than 300 ng/dL prior to surgery, with a median increase of 146 ng/dL postoperatively. This increase in testosterone level was significantly higher compared to the group with normal preoperative serum testosterone (median postoperative increase of 104 ng/dL) [44].

Similar changes in testosterone levels have not been demonstrated after surgery for benign prostate conditions, such as simple prostatectomy or transurethral resection of the prostate (TURP) [43–45]. Madersbacher et al. compared serum hormone levels in patients with prostate cancer who underwent radical prostatectomy and patients with symptomatic BPH who either underwent a TURP or were treated medically. Baseline LH and FSH levels were significantly decreased in prostate cancer patients compared to controls. Although baseline testosterone levels were also decreased, this did not reach statistical significance compared to the other two groups. After either treatment for BPH, no change in LH, FSH, or testosterone was noted at 6 and 12 months. However, after radical prostatectomy, a significant increase in FSH, LH, and testosterone levels was observed. Furthermore, patients with high-grade prostate tumors had a significantly lower baseline testosterone and a greater incremental increase in testosterone levels after radical prostatectomy, compared to patients with low-grade tumors. In contrast, the levels for patients with low-grade prostate cancer did not show significant changes post-operatively [45].

These data suggest that there may indeed be a factor secreted by prostate cancer cells that acts via an endocrine pathway to exert a negative feedback inhibition at the hypothalamic–pituitary axis.

Testosterone Replacement and Prostate Cancer

Testosterone replacement therapy in hypogonadal men has been documented to improve symptoms of fatigue, libido, and sexual performance issues [46, 47]. However, because of the inconsistencies in the literature regarding the role of androgens in prostate cancer, there has been a long-standing uncertainty about the safety of testosterone replacement therapy and its potential effects on the growth and development of prostate cancer. The traditional belief in the causative role of testosterone in prostate cancer initiation has made testosterone replacement therapy in men with prostate cancer a contraindication. Interestingly, hypogonadal men who are treated with testosterone replacement show no increased risk of prostate cancer compared to the general population [48]. In fact, testosterone may have a protective role in hypogonadal men. In 11 studies comparing testosterone to placebo, prostate cancer was diagnosed with a similar incidence in both groups; however, patients in the placebo group were found to have a higher incidence of prostate cancer detected within 6 months [48].

There have been a few small studies within the past decade that have evaluated outcomes in men who received testosterone replacement after definitive treatment of localized prostate cancer. The results were favorable and alluded to the safety of testosterone therapy in this setting [24]. Morgentaler et al. recently published a study on the effects of testosterone replacement therapy in men with untreated prostate cancer. This group evaluated 13 patients with hypogonadism and low-risk localized prostate cancer. All patients had elected active surveillance as primary management of their prostate cancer. Patients received testosterone therapy for a median of 2.5 years and the expected increase in serum testosterone levels was observed. After evaluating follow-up prostate biopsies, PSA, and serum testosterone levels, they did not find any definitive change in PSA or prostate volume and no patients had prostate cancer progression or metastatic disease during the follow-up period. Although this was a small, short-term, retrospective study, it was the first to demonstrate the potential safety of testosterone replacement in such patients [24]. The results are salient, but require further validation with larger prospective clinical trials.

Conclusion

For more than half a century, the research surrounding the relationship between androgens and prostate cancer has stemmed from the theory that testosterone administration causes prostate cancer growth. There exists a relationship between the two, but continued developments in the nature and complexity of this relationship have put to rest our traditional beliefs.

The association still remains ill defined and clinical implications are difficult to decipher. While there is mounting evidence regarding serum androgen levels, data on the relationship between tissue androgens and prostate cancer are limited. The results in contemporary literature have led to two new theories to explain this relationship: the saturation and suppression models. These concepts have far reaching implications for prostate carcinogenesis, treatment of prostate cancer, and for men suffering from concurrent hypogonadism. The role of androgens in the development and progression of prostate cancer is an ongoing controversy and these initial findings require further validation. Future research must consider longitudinal exposure to androgen, tissue concentrations of androgens, and their relationship to prostate cancer.

References

1. Jemal A, Bray R, Center M, et al. Global cancer statistics. CA Cancer J Clin. 2011;61(2):69–90.
2. Basu S, Tindall DJ. Androgen action in prostate cancer. Horm Cancer. 2010;1:223–8.
3. Cunha GR, Hayward SW, Wang YZ, et al. Role of the stromal microenvironment in carcinogenesis of the prostate. Int J Cancer. 2003;107:1–10.
4. Veltri R, Rodriguez R. Molecular biology, endocrinology, and physiology of the prostate and seminal vesicles. In: Wein AJ, Kavoussi LR, Novick AC, et al., editors. Campbell-Walsh urology. 9th ed. Philadelphia, PA: Saunders Elsevier; 2007. p. 2677–726.
5. Litvinov IV, De Marzo AM, Isaacs JT. Is the Achilles' heel for prostate cancer therapy a gain of function in androgen receptor signaling? J Clin Endocrinol Metab. 2003;88:2972–82.
6. English HF, Santen RJ, Isaacs JT. Response of glandular versus basal rat ventral prostatic epithelial cells to androgen withdrawal and replacement. Prostate. 1987;11:229–42.
7. Dehm SM, Tindall DJ. Molecular regulation of androgen action in prostate cancer. J Cell Biochem. 2006;99:333–44.
8. Labrie F, Luu-The V, Labrie C, et al. DHEA and its transformation into androgens and estrogens in peripheral target tissues: intracrinology. Front Neuroendocrinol. 2001;22:185–212.
9. Amory JK, Bremner WJ. Regulation of testicular function in men: implications for male hormonal contraceptive development. J Steroid Biochem Mol Biol. 2003;85:357–61.
10. Kaarbo M, Klokk TI, Saatcioglu F. Androgen signaling and its interactions with other signaling pathways in prostate cancer. Bioessays. 2007;29(12):1227–38.
11. Zhou ZX, Lane MV, Kemppainen JA. Specificity of ligand-dependent androgen receptor stabilization: receptor domain interactions influence ligand dissociation and receptor stability. Mol Endocrinol. 1995;9:208–18.
12. Heinlein CA, Chang C. Androgen receptor in prostate cancer. Endocr Rev. 2004;25:276–308.
13. Tindall DJ, Rittmaster RS. The rationale for inhibiting 5alpha-reductase isoenzymes in the prevention and treatment of prostate cancer. J Urol. 2008;179(4):1235–42.
14. Steers WD. 5alpha-reductase activity in the prostate. Urology. 2001;58:17–24. discussion 24.
15. Page ST, Hirano L, Gilchriest J. Dutasteride reduces prostate size and prostate specific antigen in older hypogonadal men with benign prostatic hyperplasia undergoing testosterone replacement therapy. J Urol. 2011;186(1):191–7.
16. Amory JK, Wang C, Swerdloff RS, et al. The effect of 5alpha-reductase inhibition with dutasteride and finasteride on semen parameters and serum hormones in healthy men. J Clin Endocrinol Metabol. 2007;92(5):1659–65.

17. Thompson IM, Goodman PJ, Tangen CM, et al. The influence of finasteride on the development of prostate cancer. N Engl J Med. 2003;349:215–24.
18. Andriole GL, Bostwick DG, Brawley OW, et al. Effect of dutasteride on the risk of prostate cancer. N Engl J Med. 2010;362:1192–202.
19. Kaaks R, Lukanova A, Sommersberg B. Plasma androgens, IGF-1, body size, and prostate cancer risk: a synthetic review. Prostate Cancer Prostatic Dis. 2000;3(3):157–72.
20. Huggins C, Hodges CV. Studies on prostatic cancer. I. The effect of castration, of estrogen and androgen injection on serum phosphatases in metastatic carcinoma of the prostate. Cancer Res. 1941;1:293–7.
21. Prout GR, Brewer WR. Response of men with advanced prostatic carcinoma to exogenous administration of testosterone. Cancer. 1967;20:1871.
22. Fowler JE, Whitmore WF. The response of metastatic adenocarcinoma of the prostate to exogenous testosterone. J Urol. 1981;126:372–5.
23. Pollard MM, Luckert PPH, Schmidt MMA. Induction of prostate adenocarcinomas in Lobund Wistar rats by testosterone. Prostate. 1982;3:563–8.
24. Morgentaler A, Lipshultz LI, Bennett R, et al. Testosterone therapy in men with untreated prostate cancer. J Urol. 2011;185:1256–61.
25. Gann P, Hennekens C, Ma J, et al. Prospective study of sex hormone levels and risk of prostatic carcinoma. J Natl Cancer Inst. 1996;88:1118–26.
26. Parsons JK, Carter HB, Platz EA, et al. Serum testosterone and the risk of prostate carcinoma: potential implications for testosterone therapy. Cancer Epidemiol Biomarkers Prev. 2005;14:2257–60.
27. Platz EA, Giovannucci E. The epidemiology of sex steroid hormones and their signaling and metabolic pathways in the etiology of prostate cancer. J Steroid Biochem Mol Biol. 2004;92:237–53.
28. Roddam AW, Allen NE, Appleby P, et al. Endogenous sex hormones and prostate cancer: a collaborative analysis of 18 prospective studies. J Natl Cancer Inst. 2008;100(3):170–83.
29. Pierorazio PM, Ferrucci L, Kettermann A, et al. Serum testosterone is associated with aggressive prostate cancer in older men: results from the Baltimore Longitudinal Study of Aging. BJU Int. 2009;105:824–9.
30. Statin P, Lumme S, Tenkanen L, et al. High levels of circulating testosterone are not associated with increased prostate cancer risk: a pooled prospective study. Int J Cancer. 2004;108:418–24.
31. Morgentaler A. Testosterone and prostate cancer: an historical perspective on a modern myth. Eur Urol. 2006;50(5):935–9.
32. Morgentaler A. Testosterone deficiency and prostate cancer: emerging recognition of an important and troubling relationship. Eur Urol. 2007;52:623–5.
33. Morgentaler A, Rhoden EL. Prevalence of prostate cancer among hypogonadal men with prostate-specific antigen of 4.0 ng/ml or less. Urology. 2006;68:1263–7.
34. Hoffman MA, DeWold WC, Morgentaler A. Is low serum free testosterone a marker for high grade prostate cancer? J Urol. 2000;163:824–7.
35. Yano M, Imamoto T, Suzuki H, et al. The clinical potential of pretreatment serum testosteone level to improve the efficiency of prostate cancer screening. Eur Urol. 2007;51:293–5.
36. Schatzl G, Madersbacher S, Thurridl T, et al. High-grade prostate cancer is associated with low serum testosterone levels. Prostate. 2001;47:52–8.
37. Zhang PL, Rosen S, Veeramachaneni R, et al. Association between prostate cancer and serum testosterone levels. Prostate. 2002;53(3):179–82.
38. Massengill JC, Sun L, Moul JW, et al. Pretreatment total testosterone level predicts pathological stage in patients with localized prostate cancer treated with radical prostatectomy. J Urol. 2003;169:1670–5.
39. Nishiyama T, Ikarashi T, Hashimoto Y, et al. Association between the dihydrotestosterone level in the prostate and prostate cancer aggressiveness using the Gleason score. J Urol. 2006;176:1387–91.
40. Bologna M, Muzi P, Biordi L, et al. Finasteride dose-dependently reduces de proliferation rate of the LNCAP human prostatic cancer cell line in vitro. Urology. 1995;45:282–90.

41. Morgentaler A, Traish AM. Shifting the paradigm of testosterone and prostate cancer: the saturation model and the limits of androgen-dependent growth. Eur Urol. 2009;55:310–21.
42. Marks LS, Mazer NA, Mostaghel E, et al. Effect of testosterone replacement therapy on prostate issue in men with late-onset hypogonadism: a randomized controlled trial. JAMA. 2006;296:2351–61.
43. Miller LR, Partin AW, Chan DW, et al. Influence of radical prostatectomy on serum hormone levels. J Urol. 1998;160:449–53.
44. Yamamoto S, Yonese J, Kawakami S, et al. Preoperative serum testosterone level as an independent predictor of treatment failure following radical prostatectomy. Eur Urol. 2007;52:696–701.
45. Madersbacher S, Schatzl G, Bieglmayer C, et al. Impact of radical prostatectomy and TURP on the hypothalamic-pituitary-gonadal hormone axis. Urology. 2002;60:869–74.
46. Khera M, Bhattacharya RK, Blick G, et al. Improved sexual function with testosterone replacement therapy in hypogonadal men: real-world data from the Testim registry in the United States. J Sex Med. 2011;8(11):3204–13.
47. Bhasin S, Cunningham GR, Hayes FJ, et al. Testosterone therapy in adult men with androgen deficiency syndromes: an endocrine society clinical practice guideline. J Clin Endocrinol Metab. 2006;91:1995.
48. Shabsigh R, Crawford ED, Nehra A, et al. Testosterone therapy in hypogonadal men and potential prostate cancer risk: a systematic review. Int J Impot Res. 2009;21:9–23.

Chapter 7
Androgen Deficiency in the Adolescent Male

Erin R. McNamara and Sherry S. Ross

Introduction

Puberty is an acceleration in the development of physical, psychological, and emotional factors, mainly identified by the physical changes that accompany advancement in chronologic age. When children do not undergo normal progression to puberty, both physical and psychological challenges result in social rejection and decrease in the quality of life. For this reason, androgen replacement therapy needs to be initiated with a goal of normal pubertal development and growth, improvement in psychosocial function, and an overall feeling of acceptance. The use of androgen replacement in adolescent males promotes linear growth, sexual secondary characteristics, and psychosocial development in boys who are behind their peers in development [1]. The two most common conditions where androgen replacement may be used is (1) constitutional delay of growth and puberty (CDGP) or transient hypogonadotropic hypogonadism and (2) permanent hypogonadism due to either primary or secondary hypogonadism. In this chapter, we will further discuss the physiology of puberty in males, the indications for treatment with androgen replacement therapy, how to administer testosterone, and treatment goals and anticipated effects.

E.R. McNamara, BA, MS, MD
Department of Surgery/Division of Urology, Duke University Medical Center, Durham, NC, USA

S.S. Ross, MD (✉)
Department of Surgery, Duke University Medical Center, Box 3831, Durham, NC 27710, USA
e-mail: Sherry.Ross@Duke.edu

W.J.G. Hellstrom (ed.), *Androgen Deficiency and Testosterone Replacement: Current Controversies and Strategies*, Current Clinical Urology,
DOI 10.1007/978-1-62703-179-0_7,

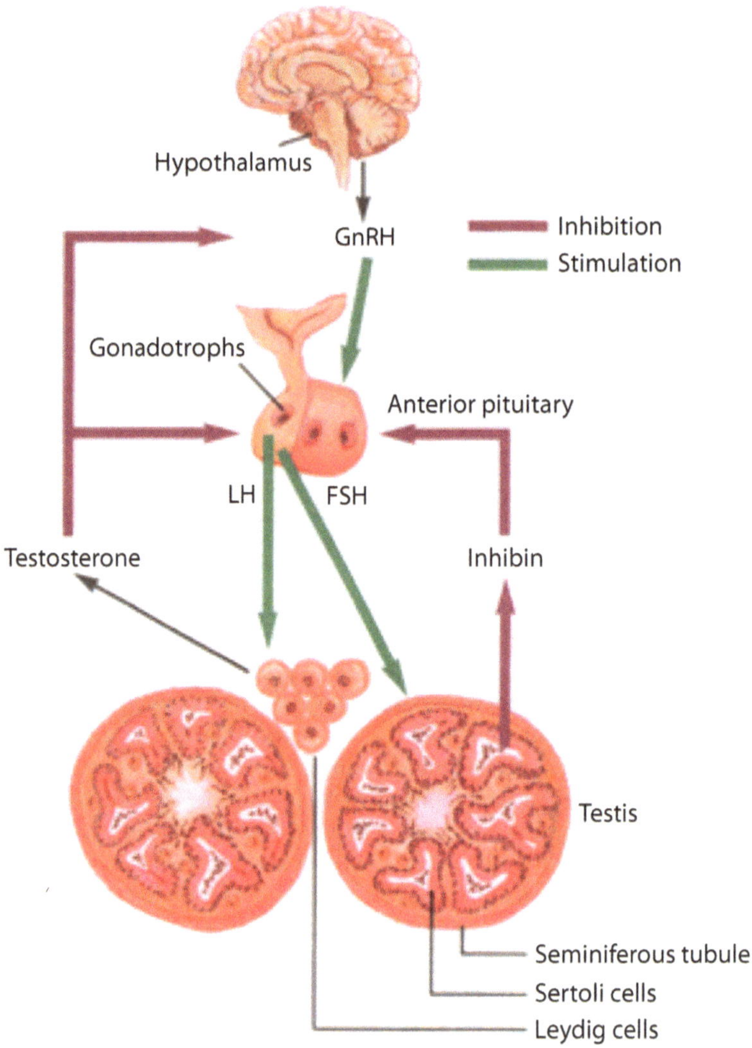

Fig. 7.1 Hypothalamus-pituitary-gonadal axis. (Reprinted from Zavos, P. (2009). Male Infertility. Retrieved from http://www.homefertility.com/mi.html. With permission from Prof. Dr. Panayiotis Zavos.)

Physiology of Puberty

Male puberty is under the control of the hypothalamic-pituitary-gonadal axis which is responsible for an accelerated increase in gonadotropins and sex steroid production (Fig. 7.1). This axis matures very early in the fetal period. During gestation, there is an increase in gonadotropin releasing hormone (GnRH) which reaches a peak at 34–38 weeks [2]. Luteinizing hormone (LH) and follicle-stimulating hormone

(FSH) secretion increase during the first month of life due to the removal of maternal estrogens, creating a testosterone surge. However, by age 6 months, GnRH and LH, FSH and subsequently, testosterone levels are low. These gonadotropins remain low during the prepubertal stage of development. At the beginning of puberty, nocturnal LH levels increase, which increases plasma testosterone during the morning hours. As this cycle continues, Ledyig cells and Sertoli cells proliferate, germ cells increase in number, and 95% of testosterone is produced in the testes [3]. The remaining 5% of sex hormones are made in the adrenal glands. While the entire chain of events leading to puberty has not been completely elucidated, investigators theorize that there is central inhibition of the HPG axis. Once the central inhibition is removed GnRH is released, LH increases, stimulating a second testosterone surge and normal pubertal development occurs [4].

The onset of normal puberty varies between children, but changes typically associated with normal development can be seen between the ages of 9–13.5 years. The mean age in western countries is 11.5 years [5]. Recent studies have suggested that puberty is occurring earlier in girls and that similarly there may be earlier genital growth in boys [6, 7]. However, these studies have been criticized for their methodology, causing concerns about the conclusions of these studies. Other investigators have found no evidence for earlier pubertal development [8, 9], leaving this question virtually unanswered.

Tanner staging defines five stages of genital and pubic hair development (Fig. 7.2). Signs of puberty in the male child include increases in testicular volume and penile girth and length. Testicular volume increases from 1–2 to 3–8 mL and finally to 20–30 mL in adulthood. As testicular volume increases, testosterone levels rise. When testicular volume reaches 4 mL, plasma testosterone increases from less than 0.1 ng/mL to over 6–7 mg of testosterone being secreted each day [10]. Increases in testosterone, stimulates penile growth. Penile size is usually measured in the flaccid stretched state. While penile size varies, the average length in early development measures 6.7 cm, which increases to 12.4+2.7 cm in the adult white male [11]. Other important aspects of puberty include growth of muscle mass, decrease in body fat and increase in bone mineral content, and skeletal muscle maturation [12].

Delayed Puberty

When expected milestones in pubertal development are not met, this raises concerns for pubertal delay. Typically, the first step after excluding obvious infirmities is a full endocrinology evaluation. While most parents and children often seek out medical attention earlier, delayed puberty is diagnosed when a male child does not undergo full pubertal development by age of 16 or 17 [13].

One of the most common reasons for delay in normal pubertal development is termed CDGP or transient delayed puberty. Often, families report a known "late bloomer" in the family tree. Parents seek treatment due to delayed growth and delay

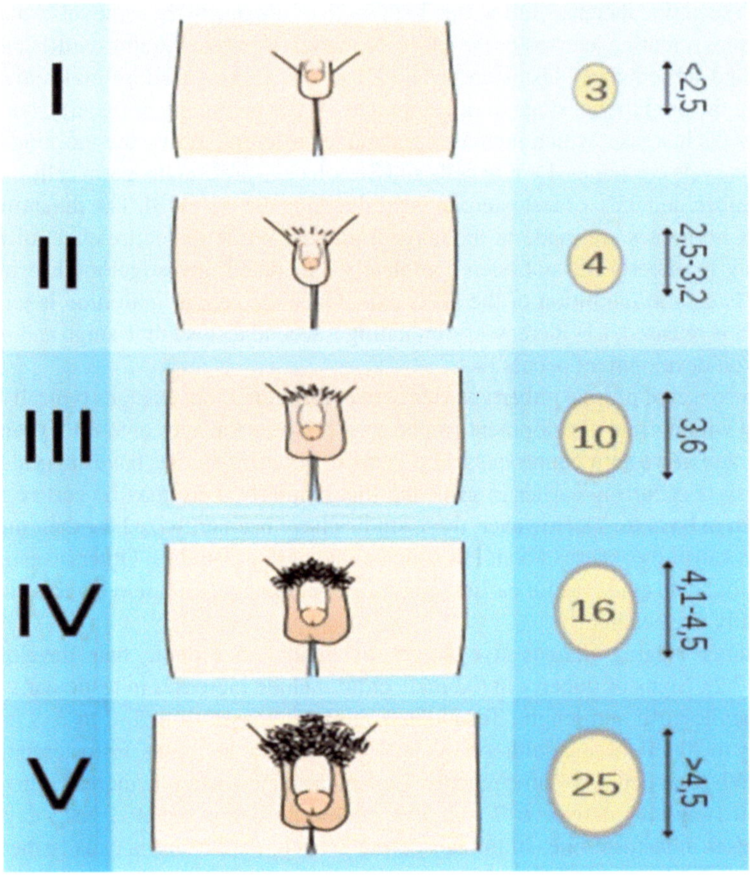

Fig. 7.2 Male tanner stage. (Reprinted from Tanner Scale. http://en.wikipedia.org/wiki/File:Tanner_scale-male.svg. Last Accessed April 24th, 2012.)

in the tempo of puberty coupled with delayed psychosocial development as it relates to the child's peers [14]. While physical and psychosocial development does not necessarily represent chronologic age, these boys are appropriately developed for their bone age. This means that bone age is delayed behind chronological age and the developmental milestones are reached only at a normal bone age. The social implications of delayed puberty for these boys are daunting. The Oakland Growth study followed 16 early and 16 late-maturing males from age of 11 through adulthood. This study found that peers considered delayed males less attractive, more talkative, and less mature. The affected male child describes feelings of inadequacy and rejection through adolescence and into adulthood [15]. A second large study conducted by the National Health Examinations Survey found that teachers usually rated the males affected with delayed puberty as less intelligent. These children had lower intellectual test scores and had lower parental expectations of academic achievement [16]. In addition to the psychosocial issues, there are also physical impediments to delayed puberty. Studies have shown that without treatment, these children do not reach predicted adult height and bone mineral density may be diminished [17, 18]. Androgen therapy in children with CDGP can have beneficial effects on bone mass, body composition, and important psychological aspects. Testosterone is used short-term until the HPG axis catches up and allows for normal spontaneous pubertal development.

Primary or secondary hypogonadism can also cause delayed puberty. This is a permanent form of hypogonadism and requires lifelong androgen replacement therapy. The absence of hormones may be due to hypogonadotropic hypogonadism. In hypogonadotropic hypogonadism, there is insufficient pulsatile GnRH secretion and subsequently a LH/FSH deficiency. The etiology of hypogonadotropic hypogonadism can include tumors, inflammatory processes, trauma, genetic conditions, and developmental defects. Children that present with neurologic signs, anosmia (associated with Kallmann syndrome), headaches, or visual disturbances should be evaluated more thoroughly searching for a central nervous system lesion and have a full hormonal profile. Other causes include anorexia nervosa, malnutrition, and very competitive athletic training. Investigators theorize that in these conditions, an increase in cortisol releasing factor increases beta-endorphins and inhibits GnRH. Increased stress inhibits the HPG axis, resulting in delayed puberty in these males [19]. It can be difficult at times to differentiate between CDGP and permanent hypogonadotropic hypogonadism. However, most experts do not support special testing if the only complaint is delay in pubertal development. If needed, these entities can usually be distinguished once androgen therapy is initiated [1]. Children with permanent hypogonadism have a hypothalamic/pituitary gonadotropin deficiency and while GnRH can be administered, the use for pulsatile administration is somewhat impractical. Testosterone supplementation is a simple solution, and these males will require lifelong androgen therapy, since the normal HPG axis is nonfunctional.

Another rare cause of permanent hypogonadism is impaired secretion of gonadal steroids leading to decreased negative feedback of the hypothalamus. In this case, there is low plasma testosterone in the presence of high plasma LH and FSH levels. A common genetic abnormality associated with hypergonadotropic hypogonadism is Klinefelter syndrome, in which case a 47,XXY karyotype or a mosaic 46,XY/47,XXY is most common [20]. In infancy and childhood, the symptoms associated with Klinefelter syndrome are often nonspecific and the diagnosis can be missed. During adolescence, puberty often starts normally, corresponding to the patient's peer group, with genital enlargement and pubic hair growth. However, while the testes start to grow, they typically skip growing at 6 mL, leaving a discordance between the degree of sexual development and the size of the testes which often leads to the diagnosis [21]. In adulthood, small testes and a bivariate testosterone/LH evaluation outside the normal reference range is a tell-tale sign of Klinefelter syndrome. As the Klinefelter patient moves through adolescence and into adulthood, frequent plasma testosterone and LH monitoring is necessary. In some cases, testosterone replacement is considered when the LH is >2.5 SD above the mean value of 2–14 U/L or when the plasma testosterone is less than 84–480 ng/dL [22]. In many cases, testosterone replacement

Hypogonadism	Transitory delayed puberty
Hypergonadotropic hypogonadism Klinefelter syndrome[a] and variants (e.g. XX males, 48, XXYY, etc.) Congenital absence of testes (anorchia) LH resistance Biosynthetic defects of testosterone Androgen insensitivity Iatrogenic	*Primary* constitutional delay of puberty
Hypogonadotropic hypogonadism Gonadotropin deficiency (congenital[b] or acquired[c]; isolated or associated with anosmia or multiple pituitary deficiencies) Isolated LH deficiency Genetic syndromes (e.g. Noonan syndrome, Prader-Willi syndrome) Chronic diseases	*Secondary to:* chronic disorders nutritional imbalance (including anorexia nervosa and related disorders) chronic endocrine diseases

[a] Many patients with Klinefelter syndrome do not start puberty late, but they may require androgen supplementation to complete pubertal development. [b] Many genes may be causative. [c] Neoplasia, traumatic injury, irradiation, etc.

Fig. 7.3 Causes of delayed puberty in males. (Based on data from Bertelloni et al. [26], Kaplowitz, P.B., Delayed puberty. Pediatr Rev, 2010. 31(5): p. 189–95, Leichtnam, M.L., et al., Testosterone hormone replacement therapy: state-of-the-art and emerging technologies. Pharm Res, 2006. 23(6): p. 1117–32.)

may be unnecessary, unless an infertility evaluation is initiated. Figure 7.3 lists many of the causes of delayed puberty in the male adolescent.

Androgen Replacement

The primary goals of androgen replacement therapy in the adolescent male is to promote the linear growth, secondary sexual characteristics, accrual of adequate bone mineral content, and acquisition of normal muscle mass [1, 21]. Secondary goals include normal psychosocial function. In order to obtain these goals, the physician needs to recognize the tenets of testosterone replacement; first of which is to replace testosterone only in those males who are hypogonadal and to insure that the testosterone level is brought to a normal physiologic state. These goals should be obtained regardless of the etiology of hypogonadism. Testosterone replacement must be safe and effective with minimal side effects. While many forms of testosterone replacement have been clinically used in the adult population and are efficacious, these regimens cannot be applied to the adolescent, since a full adult dose of testosterone is not required to initiate or maintain puberty in this age group.

Testosterone formulation	Drug	Starting dose	Optimal adult dose	Formulation-specific advantages	Formulation-specific disadvantages
Injectable (i.m.)	testosterone enanthate	25–50 mg/2–4 weeks	200–250 mg/ 2–3 weeks	large clinical experience in adolescence	peaks and troughs in circulating testosterone, gynecomastia, local pain, mood disturbances
	testosterone cypionate	similar to testosterone enanthate	200 mg/2 weeks		
	testosterone undecanoate	–	1,000 mg/12 weeks (range 10–14 weeks)	stable serum testosterone levels	local pain; lack of experience in adolescence
Oral	testosterone undecanoate[a]	40 mg/day × 6–12 months [16]	40–80 mg × 2–3 times/day	oral administration	variable clinical effects, fluctuating hormone levels
Trans-dermal	scrotal patch	–	4–6 mg/day	mimics circadian variations	skin irritation, shaving of the scrotum, abnormal high DHT levels
	non scrotal patch[b]	14–16 years: 2.5 mg/ 12 night h [19] 17–19 years: 2.5 mg/day [19] >20 years: 5.0 mg/day [19] 12.5–15.0 years: 5 mg/8–12 h [20]	2.5–5.0 mg/day	mimics circadian variations	skin irritation; little experience in adolescence
	gel 1%	0.5 g/day, increasing dose based on testosterone levels [21]	5–10 mg/day	mimics circadian variations, good clinical response, no visible patch, gel dries quickly	potential transfer to other people; little experience in adolescence
	gel 2% (puff 0.5 mg)	–	2–4 mg/day	as 1% gel; possibility of individualized doses	potential transfer to other people; no experience in adolescence
Transbuccal	biopellet 30 mg	–	1–2 cps/day	absorption directly into systemic circulation, stable serum levels, no visible patch	taste alteration, gum irritation, no experience in adolescence

Fig. 7.4 Preparations of testosterone. (Based on data from Bhasin et al. [23], Bertelloni et al. [26], Leichtnam, M.L., et al., Testosterone hormone replacement therapy: state-of-the-art and emerging technologies. Pharm Res, 2006. 23(6): p. 1117–32.)

Treatment

Testosterone was first synthesized in the 1930s. The original oral form of T underwent rapid liver metabolism and therefore, was not useful in long-term management of hypogonadism. Other forms were subsequently introduced, but due to hepatotoxicity, were deemed dangerous for human use. The most successful forms of therapy have been the esters (short and long acting). These compounds can be injected and are lipophilic and very durable. More recently, the addition of transdermal forms of testosterone and pellets have made administration of testosterone more convenient for the affected patient. Figure 7.4 gives a brief description of the different forms of testosterone in clinical practice. Note that not all forms of T are available in the US. Several studies have evaluated different testosterone formulations in male adolescents [23–26]. Testosterone enanthate is the gold standard, as it increases muscle mass, accelerates linear bone growth, and induces the appearance of secondary sexual characteristics. This allows for accomplishing the goals of androgen therapy in adolescents [27–29]. Prior to the initiation of androgen replacement therapy, the child presenting with delayed puberty should undergo a complete history and physical examination. Any additional abnormalities need to be addressed first. A complete bone age evaluation including bone density scans of lumbar spine and femoral heads is obtained. When the timing for the therapy is appropriate, testosterone enanthate or cypionate is initiated at a starting dose of

50–100 mg IM every 4 weeks for 3 months. At the 3-month mark, most patients will notice increases in appetite, body weight, and height. There may also be an increase in testicular size. An early morning testosterone sample is obtained 3 weeks after injection to measure endogenous levels. If no physical changes are noted, the dose can be increased by 25–50 mg IM every 4 weeks for another 3 months. Once the testicular volume reaches approximately 10 mL, exogenous testosterone can be discontinued and the HPG axis takes over and allows puberty to continue. If there is no increase in testicular size or no response after a 1 to 2 years of treatment, permanent hypogonadism is diagnosed.

For adolescents with permanent hypogonadism, treatment is continued for life. A gradual increase in the frequency of testosterone injections is indicated in order to mimic adult testosterone levels. The testosterone is gradually increased to 200 mg every 2–3 weeks. When adult levels are reached, it may be convenient to switch to transdermal or pellet preparations of testosterone.

Monitoring of androgen replacement therapy in adolescents requires regular follow up visits with the physician to evaluate growth, pubertal changes, and for bone age assessment. These visits should include laboratory testing of testosterone, most often timed just before the next treatment injection to adequately evaluate endogenous testosterone levels. A complete blood count, including hematocrit, needs to be ruled out for polycythemia (>54%), which is a contraindication to continuing therapy unless appropriately addressed.

The side effects of testosterone replacement include disproportionately advancing skeletal maturation, a rapid change in libido, priapism, polycythemia, acne, gynecomastia, and obstructive sleep apnea. Patients and parents need to be adequately informed about these potential consequences. While rare, these changes could be detrimental to the patient and should be monitored with each physician visit [30]. When normal testosterone levels are attained, side effects are usually minimal. Transdermal and pellet preparations are difficult to obtain and maintain a steady state of plasma testosterone in the adolescent and should not be used for initiation of therapy. Once the adolescent has reached his full adult height and progressed through puberty, he can be switched to other more convenient forms of testosterone if he has permanent hypogonadism.

Conclusion

It is imperative to keep the goals of androgen replacement therapy in the adolescent in focus. For the majority of boys with delayed puberty, this is a stepping-stone from prepubertal hypogonadism to normal adolescence. Androgen replacement therapy has been shown to have significant physical and psychological benefits that help with social acceptance by peers and feeling of well-being for these adolescents who are maturing into adulthood. The same holds true for males with permanent hypogonadism, with the exception that lifelong androgen replacement therapy is required. In the adolescent, treatment typically begins around age 12–13 years and is optimally done with the injectable testosterone esters that make management of plasma testosterone levels easier to monitor and adjust. While side effects are rare, close monitoring is important during testosterone therapy to avoid supraphysiologic levels. Families, patients, and physicians all play important roles in the use of androgen therapy in the adolescent male with delayed puberty.

References

1. Rogol AD. New facets of androgen replacement therapy during childhood and adolescence. Expert Opin Pharmacother. 2005;6(8):1319–36.
2. Siler-Khodr TM, Khodr GS. Studies in human fetal endocrinology. I. Luteinizing hormone-releasing factor content of the hypothalamus. Am J Obstet Gynecol. 1978;130(7):795–800.
3. Dunkel L, et al. Gonadal control of pulsatile secretion of luteinizing hormone and follicle-stimulating hormone in prepubertal boys evaluated by ultrasensitive time-resolved immunofluorometric assays. J Clin Endocrinol Metab. 1990;70(1):107–14.
4. Plant TM. Neurobiological bases underlying the control of the onset of puberty in the rhesus monkey: a representative higher primate. Front Neuroendocrinol. 2001;22(2):107–39.
5. Knorr D, et al. Plasma testosterone in male puberty. I. Physiology of plasma testosterone. Acta Endocrinol (Copenh). 1974;75(1):181–94.
6. Kaplowitz PB, Oberfield SE. Reexamination of the age limit for defining when puberty is precocious in girls in the United States: implications for evaluation and treatment. Drug and Therapeutics and Executive Committees of the Lawson Wilkins Pediatric Endocrine Society. Pediatrics. 1999;104(4 Pt 1):936–41.
7. Herman-Giddens ME, Wang L, Koch G. Secondary sexual characteristics in boys: estimates from the national health and nutrition examination survey III, 1988–1994. Arch Pediatr Adolesc Med. 2001;155(9):1022–8.
8. Biro FM, et al. Pubertal staging in boys. J Pediatr. 1995;127(1):100–2.
9. de Muinich Keizer SM, Mul D. Trends in pubertal development in Europe. Hum Reprod Update. 2001;7(3):287–91.
10. August GP, Grumbach MM, Kaplan SL. Hormonal changes in puberty. 3. Correlation of plasma testosterone, LH, FSH, testicular size, and bone age with male pubertal development. J Clin Endocrinol Metab. 1972;34(2):319–26.
11. Sutherland RS, et al. The effect of prepubertal androgen exposure on adult penile length. J Urol. 1996;156(2 Pt 2):783–7. discussion 787.
12. Veldhuis JD, et al. Endocrine control of body composition in infancy, childhood, and puberty. Endocr Rev. 2005;26(1):114–46.
13. Styne DM, Grumbach MM. Puberty: ontogeny, neuroendocrinology, physiology and disorders. Williams textbook of endocrinology. 11th edition. Ed. Kronenberg HM, Melmed S, Polonsky KS, Larsen PR. Philadelphia: Saunders Elsevier; 2008.
14. Keenan BS, et al. Androgen-stimulated pubertal growth: the effects of testosterone and dihydrotestosterone on growth hormone and insulin-like growth factor-I in the treatment of short stature and delayed puberty. J Clin Endocrinol Metab. 1993;76(4):996–1001.
15. Gross RT, Duke PM. The effect of early versus late physical maturation on adolescent behavior. Pediatr Clin North Am. 1980;27(1):71–7.
16. Duke PM, et al. Educational correlates of early and late sexual maturation in adolescence. J Pediatr. 1982;100(4):633–7.
17. Albanese A, Stanhope R. Predictive factors in the determination of final height in boys with constitutional delay of growth and puberty. J Pediatr. 1995;126(4):545–50.
18. Houchin LD, Rogol AD. Androgen replacement in children with constitutional delay of puberty: the case for aggressive therapy. Baillieres Clin Endocrinol Metab. 1998;12(3):427–40.
19. Opstad PK. Androgenic hormones during prolonged physical stress, sleep, and energy deficiency. J Clin Endocrinol Metab. 1992;74(5):1176–83.

20. Paulsen CA, et al. Klinefelter's syndrome and its variants: a hormonal and chromosomal study. Recent Prog Horm Res. 1968;24:321–63.
21. Rogol AD, Tartaglia N. Considerations for androgen therapy in children and adolescents with Klinefelter syndrome (47, XXY). Pediatr Endocrinol Rev. 2010;8 Suppl 1:145–50.
22. Nielsen J, Pelsen B, Sorensen K. Follow-up of 30 Klinefelter males treated with testosterone. Clin Genet. 1988;33(4):262–9.
23. Bhasin S, et al. Testosterone therapy in adult men with androgen deficiency syndromes: an endocrine society clinical practice guideline. J Clin Endocrinol Metab. 2006;91(6):1995–2010.
24. Lenzi A, et al. Epidemiology, diagnosis, and treatment of male hypogonadotropic hypogonadism. J Endocrinol Invest. 2009;32(11):934–8.
25. Richmond EJ, Rogol AD. Male pubertal development and the role of androgen therapy. Nat Clin Pract Endocrinol Metab. 2007;3(4):338–44.
26. Bertelloni S, et al. Androgen therapy in hypogonadal adolescent males. Horm Res Paediatr. 2010;74(4):292–6.
27. Rosenfeld RG, Northcraft GB, Hintz RL. A prospective, randomized study of testosterone treatment of constitutional delay of growth and development in male adolescents. Pediatrics. 1982;69(6):681–7.
28. Richman RA, Kirsch LR. Testosterone treatment in adolescent boys with constitutional delay in growth and development. N Engl J Med. 1988;319(24):1563–7.
29. Bergada I, Bergada C. Long term treatment with low dose testosterone in constitutional delay of growth and puberty: effect on bone age maturation and pubertal progression. J Pediatr Endocrinol Metab. 1995;8(2):117–22.
30. Drobac S, et al. A workshop on pubertal hormone replacement options in the United States. J Pediatr Endocrinol Metab. 2006;19(1):55–64.

Chapter 8
Effects of Androgen Deficiency and Replacement on Male Fertility

Andrew C. Kramer

Introduction

Androgen deficiency may manifest at various periods in a man's life. Its etiology is varied. It is particularly challenging to manage this situation in a man who desires future fertility. Many of the therapeutic options for androgen replacement require exogenous administration of testosterone, which can suppress spermatogenesis. In young patients who may desire fertility in the near future or who are actively trying to conceive, the gamut of options for androgen replacement must be considered carefully.

Mechanism

Hormonal control of spermatogenesis is based on the interaction between the hypothalamus, pituitary, and testicles. High levels of testosterone, up to 20–50 times serum levels, are needed locally in the testicle to sustain spermatogenesis, indicating that this process is under paracrine control [1–3]. The exact mechanism of hormonal control is outlined elsewhere. Briefly, in response to gonadotropin-releasing hormone (GnRH), the hypothalamus stimulates the anterior pituitary to release luteinizing hormone (LH) and follicle-stimulating hormone (FSH) [4].

Leydig cells, located in the interstitial spaces of the testis, respond to LH and release androgens (19-carbon steroids) [5]. These include testosterone, androstenedione, and dehydroepiandrosterone [6]. FSH acts on Sertoli cells, also located in the testis, to produce androgen-binding protein (ABP), which is critical for concentrating

A.C. Kramer, MD (✉)
Department of Surgery, The University of Maryland School of Medicine,
29 Greene Street, Suite 500, Baltimore, MA 21201, USA
e-mail: Akramer@smail.umaryland.edu

W.J.G. Hellstrom (ed.), *Androgen Deficiency and Testosterone Replacement: Current Controversies and Strategies*, Current Clinical Urology,
DOI 10.1007/978-1-62703-179-0_8, © Springer Science+Business Media, LLC 2013

testosterone at high levels necessary to maintain spermatogenesis [7]. ABP concentrates testosterone in the immediate proximity of the developing germ cells.

Pulsatile secretion of GnRH from the hypothalamus is fundamental to the functionality of this pathway [8]. Failure of this axis to work in an episodic fashion results in hypogonadism.

Etiologies of Androgen Deficiency

The differential diagnosis of hypogonadotropic hypogonadism (HH) is extensive and can be classified as either congenital or acquired. There are forms of idiopathic HH such as Kallmann syndrome, adult onset IHH, fertile eunuch syndrome, and congenital adrenal hypoplasia [9]. There are also genetic defects involving the gonadotropic subunits that render them ineffective as well as an array of other pituitary disturbances [10]. Additionally, there are a few disorders present with obesity, such as Prader–Willi syndrome and Laurence–Moon–Bordet–Biedl syndrome [11, 12].

There are acquired forms of HH as well, ranging from tumors of the hypothalamus and pituitary along with other systemic disorders such as histiocytosis X and sarcoidosis [13]. Previous surgery, head trauma, and radiation can induce acquired HH.

The more common form of hypogonadism seen in young male patients is primary testicular failure. There is a range of etiologies here as well, ranging from idiopathic testicular failure, trauma, and undescended testes [14]. Certain conditions can also induce testicular failure such as extreme exercise, exogenous anabolic steroid use, glucocorticoid therapy, chronic diseases, and narcotics [15–17].

The Challenge to Clinicians

The difficulty in treating the hypogonadal man is that exogenous steroids negatively feedback and suppress the hypothalamic–pituitary axis. While administration of androgens solves the systemic effects of low testosterone, high local testosterone concentrations cannot be maintained. This important consideration must be addressed when treating a man with hypogonadism who wishes to have children.

Effects of Androgen Deficiency

While it's recognized that serum testosterone levels decline with aging, it's unclear what exactly the clinical symptoms of "andropause" are and whether there is a need to treat symptoms. It remains unsolved whether men derive a significant benefit from androgen replacement treatment [18].

There are numerous symptoms observed with a low androgen state, both in young men, and in the aging male. These include low libido, decrease in stamina and/or strength, reduced muscle mass and increased fat, osteopenia, and gynecomastia [18].

Little direct evidence has associated low serum testosterone levels with an increased risk of atherosclerosis and the metabolic syndrome. Other studies have documented anti-ischemic actions of androgen therapy as a potential benefit [19].

Erectile function has not been conclusively linked to normal androgen levels. While a normal testosterone level is thought to be necessary for libido and sexual thoughts and dreams, it remains unproven if sexual function can be fully optimized with androgen supplementation alone [20]. Young men may sense that testosterone therapy is central to their ability to achieve and maintain potency. This belief, whether true or not, creates the challenge for healthcare providers in treating such patients. While the goal is to maintain an acceptable androgen profile in young men, the therapeutic approach must be altered so that spermatogenesis is maintained.

Therapeutic Options

Spermatogenesis cannot be induced in men with primary hypogonadism [21], defined as testicular or seminiferous tubule failure [21]. Sperm production can be stimulated in men with secondary hypogonadism, defined as functional failure of the hypothalamus or pituitary. In such patients, it is important to stop any exogenous androgen use and begin one of the alternate therapies described below. For men taking exogenous steroids for bodybuilding, the diagnosis can be made by a good history and hormone profile. The profile usually shows normal serum androgens with markedly suppressed levels of LH and FSH, due to negative feedback [22]. This select population of men with iatrogenic testicular failure is often salvaged by ceasing the exogenous steroid use and allowing enough time for the testicle and hypothalamic–pituitary axis to recover. In most cases it does and spermatogenesis resumes.

In the following section, a rational approach for androgen replacement in men with secondary hypogonadism who desire fertility is presented. The major treatment options discussed are the following:

1. Human chorionic gonadotropin (hCG) therapy initially with hMG added after a delay
   (a) Consideration of recombinant LH
   (b) Consideration of recombinant FSH (r-hFSH)
2. Pulsatile GnRH
3. Clomiphene citrate

The following description of hormone treatment to induce fertility in secondary hypogonadism is not the only treatment regimen [23], however, it provides a reasonable algorithm for consideration and is based upon recommendation by Drs. Snyder, Matsumoto, and Martin.

Gonadotropin Therapy

In cases of secondary hypogonadism, the etiology needs to be defined, specifically if it is at the level of the hypothalamus or pituitary gland. Men with pituitary disease can be successfully treated with gonadotropin administration alone, while hypothalamic disease is treated with either gonadotropins or GnRH.

Secondary hypogonadism is defined as a lack of secretions of the gonadotropins LH and FSH. In theory, administration of these hormones alone should reverse the disorder. LH alone can produce neutral testosterone levels, but often not enough to induce spermatogenesis [24]. Within this population, a better prognosis is noted in men who have developed hypogonadism after puberty, have incomplete or partial hypogonadism, or have descended testis by the first year of life.

hCG is similar in biologic activity to LH. It functions by stimulating the Leydig cells to make and secrete testosterone. In actuality, LH is given by its proxy, hCG, since it has a longer half-life than recombinant human LH, making it easier to administer to patients.

hCG therapy is given first, prior to intitating FSH treatment. The initial hCG treatment will give an initial boosting of intratesticular testosterone, up to 100 times higher than peripheral levels. hCG alone may be enough to induce spermatogenesis, whereas FSH alone is merely supportive, and does not have a lead action. Furthermore, hCG is less expensive than exogenous FSH. Therefore, the recommended regimen involves first stopping any exogenous testosterone therapy, starting with hCG, then after a delay, starting FSH therapy.

Regimen of hCG Therapy [23]

- Patients self-administer hCG intramuscularly into the thigh at a starting dose of 2,000 units, three times per week.
- Serum testosterone is checked every month to make sure the range of testosterone is between 400 and 900 ng/DL. The final dose can vary from 500 to 10,000 units per injection.
- Semen analyses are checked periodically as soon as the testosterone level is felt to be replete. A return of spermatogenesis may take 3–6 months in some cases.
- Addition of the FSH, by its substitute hMG, is administered once testosterone has normalized for at least 3 months of the semen analysis is half of normal values for that individual. This may arbitrarily be set at 20 million motile sperm/cc.

Regimen of Human Menopausal Gonadotropin [23]

This administration contains both FSH and LH, and is commonly used to substitute for FSH in attempt to induce spermatogenesis. FSH is believed to influence Sertoli cells to produce nutrients and provide a supportive environment for sperm production within the testis.

- A first dose of 75 units is given three times per week.
- Although maximal beneficial effects in spermatogenesis may be seen at anywhere from 3–24 months, frequent semen analyses are performed to detect progress.
- hMG dose can be doubled to 150 units, if the sperm count does not reach normal levels by 6 months.
- As soon as pregnancy occurs, this hMG regimen can be stopped, mainly because of its high cost.

Recombinant FSH

Since hMG contains some LH, the recombinant form of FSH was developed for use in women for ovulation induction, simply to avoid that small amount of LH in the preparation. In men seeking fertility, the efficacy of hMG and recombinant FSH is similar. The cost is approximately twice that of the hMG, therefore it is not indicated as a first-line therapy in men to be used in conjunction with hCG.

Pulsatile GnRH [23]

Another alternative is dosing in a pulsatile fashion in men with hypothalamic disease. A pump and syringe are used to deliver GnRH in a theoretically physiologic manner, so a pulse is given approximately every 2 h. The man wears this pump until the couple achieves pregnancy, with the dose being delivered via subcutaneous needle. The dose is started as 25 ng/kg of patient body weight, but in some men, the dose is increased to 600 ng/kg of body weight. The response time is slower for this therapy, and ranges from 1 to 3 years. This is very effective in producing normal serum testosterone levels with hypogonadal patients with hypothalamic failure. Availability in the United States is limited with this therapy; it is expensive, and efficacy is similar to that of the hCG/hMG therapy discussed above.

Clomiphene Citrate

Clomiphene citrate, or Clomid, acts as a weak estrogen receptor antagonist that stimulates gonadotropin secretion. Numerous studies have tested this agent's efficacy in men attempting to increase their sperm density who have oligospermia in the setting of low or low/normal testosterone levels. Results have not been definitive in this specific population of infertile men [25]. To date, clomiphene has shown no benefit in treating men with secondary hypogonadism due to pituitary dysfunction [26, 27]. Furthermore, men with testicular failure have also not achieved increased fertility with the use of clomiphene. Of note, clomiphene will temporarily increase serum testosterone levels, but has thus far proven ineffective in the treatment of infertility in men with idiopathic infertility or primary testicular failure [28, 29]. A dose commonly used is 50 mg by mouth every other day with frequent evaluation of semen analysis and serum testosterone levels.

Assisted Reproductive Techniques

Some men may achieve normal serum testosterone levels and even acceptable sperm densities, yet this fails to yield a pregnancy. In this scenario, the risks and costs associated with hormone replacement must be weighed against the use of an assisted reproductive technique. If there is suitable sperm, either in the ejaculate or harvested by testicular sperm extraction, they can be used in conjunction with in vitro fertilization (IVF) or intracytoplasmic sperm injection (ICSI). Success rates for these techniques vary, but reported biochemical pregnancy rates are approximately 30 % per cycle [30]. Donor insemination or adoption is also an option to be considered if there exist severe impairments in the spermatogenic pathway that androgen substitution fails to solve.

References

1. Sullivan KA, Silverman A-J. The ontogeny of gonadotropin-releasing hormone neurons in the chick. Neuroendocrinology. 1993;58:597–608.
2. Spratt DI, O'Dea LSL, Schoenfeld D, Butler J, Rao PN, Crowley Jr WF. Neuroendocrine-gonadal axis in men: frequent sampling of LH, FSH, and testosterone. Am J Physiol. 1998;254:E658–66.
3. Bilinska B, Genissel C, Carreau S. Paracrine effect of seminiferous tubule factors on rat leydig cell testosterone production: role of cytoskeleton. Biol Cell. 1997;89:435–42.
4. Charlton H. Hypothalamic control of anterior pituitary function: a history. J Neuroendocrinol. 2008;20:641–6.
5. Payne A, Hardy M, Russell L. The Leydig cell. Vienna, FL: Cache River Press; 1996. p. 301–4.
6. Payne A, Hardy M. The Leydig cell in health and disease. New York: Humana Press; 2007. p. 221–4.
7. Blok LJ, Mackenbach P, Trapman J, Themmen A, Brinkman AO, Grootegoed A. Follicle stimulating hormone regulates androgen receptor mRNA in Sertoli cells. Mol Cell Endocrinol. 1989;63:267–71.
8. Jakacki RI, Kelch RP, Sauder SE, Lloyd JS, Hopwood NJ, Marshall JC. Pulsatile secretion of luteinizing hormone in children. J Clin Endocrinol Metabol. 1982;55:453–8.

9. Lieblich JM, Rogol AD, White BJ, Rosen SW. Syndrome of anosmia with hypogonadotropic hypogonadism—a treatable form of male infertility. N Engl J Med. 1997;336:410–5.
10. Spratt DI, Carr DB, Merriam GR, Scully RE, Rao PN, Crowley Jr WF. The spectrum of abnormal patterns of gonadotropin-releasing hormone secretion in men with idiopathic hypogonadotropic hypogonadism: clinical and laboratory correlations. J Clin Endocrinol Metabol. 1987;64:283–91.
11. Santoro N, Filicori M, Crowley Jr WF. Hypogonadotropic disorders in men and women: diagnosis and therapy with pulsatile gonadotropin-releasing hormone. Endocr Rev. 1986;7:11–23.
12. Schwankhaus JD, Currie J, Jaffe MJ, Rose SR, Sherins RJ. Neurologic findings in men with isolated hypogonadotropic hypogonadism. Neurology. 1989;39:223–6.
13. Goldman L, Bennett JC, editors. Cecil textbook of medicine. 21st ed. Philadelphia: W.B. Saunders; 2000.
14. Styne DM. Puberty and its disorders in boys. Endocrinol Metab Clin North Am. 1991;20:43–69.
15. Baker HW. Reproductive effects of nontesticular illness. Endocrinol Metab Clin North Am. 1998;27:831–50.
16. Abs R, Verhelst J, Maeyaert J. Endocrine consequences of long-term intrathecal administration of opioids. J Clin Endocrinol Metab. 2000;85:2215–22.
17. Van den Berghe G, de Zegher F, Lauwers P, Veldhuis JD. Luteinizing hormone secretion and hypoandrogenaemia in critically ill men: effect of dopamine. Clin Endocrinol (Oxf). 1994;41:563–9.
18. Brawer M. Testosterone replacement in men with Andropause: an overview. Rev Urol. 2004;6 Suppl 6:S9–15.
19. Rosano G, Leonardo F, Pagnotta P, Pelliccia F, Panina G, Cerquetani E, et al. Acute anti-ischemic effect of testosterone in men with coronary artery disease. Circulation. 1999;99:1666–70.
20. Rajfer J. The relationship between testosterone and erectile dysfunction. Rev Urol. 2000;2:122–8.
21. Burris AS, Rodbard HW, Winters SJ, Sherins RJ. Gonadotropin therapy in men with isolated hypogonadotropic hypogonadism: the response to human chorionic gonadotropin is predicted by initial testicular size. J Clin Endocrinol Metab. 1988;66:1144.
22. Jarow JP, Lipshultz L. Anabolic steroid induced hypogonadotropic hypogonadism. Am J Sports Med. 1990;18:429.
23. Snyder P, Matsumoto A, Martin K. Induction of fertility in men with secondary hypogonadism. In: Basow DS, editor. UpToDate. Waltham, MA: UpToDate; 2011.
24. Finkel DM, Phillips JL, Snyder PJ. Stimulation of spermatogenesis by gonadotropins in men with hypogonadotropic hypogonadism. N Engl J Med. 1985;313:651.
25. Whitten SJ, Nangia AK, Kolettis PN. Select patients with hypogonadotropic hypogonadism may respond to treatment with clomiphene citrate. Fertil Steril. 2006;86:1664.
26. Sharlip ID, Jarow JP, Belker AM, Lipshultz LI, Sigman M, Thomas AJ, et al. Best practice policies for male infertility. Fertil Steril. 2002;77:873–82.
27. Sorbie PJ, Perez-Marrero R. The use of clomiphene citrate in male infertility. J Urol. 1984;131:425–9.
28. Ribeiro RS, Abucham J. Clomiphene fails to revert hypogonadism in most male patients with conventionally treated pituitary adenomas. Arq Bras Endocrinol Metabol. 2011;55:266–71.
29. Moradi M, Moradi A, Alemi M, Ahmadnia H, Abdi H, Ahmadi A, et al. Safety and efficacy of clomiphene citrate and L-carnitine in idiopathic male infertility: a comparative study. J Urol. 2010;7:188–93.
30. Papanikolaou EG, Camus M, Kolibianakis EM, Van Landuyt L, Van Steirteghem A, Devroey P. In vitro fertilization with single blastocyst-stage versus single cleavage-stage embryos. N Engl J Med. 2006;354:1139.

Chapter 9
HIV and Testosterone in Men

Gina Cambareri, Jane Ashby, David Goldmeier, and Hossein Sadeghi-Nejad

Introduction

In recent years, in the era of antiretroviral therapy (ARV), the overall growth of the HIV epidemic has begun to plateau with a dramatic reduction in HIV-associated mortality [1, 2] and a reduction in the annual number of new HIV infections [3]. As life expectancy for those living with HIV and access to ARV increases, the total number of individuals living globally with HIV remains high and was estimated to be around 33 million by the end of 2009 [3]. As treatments for HIV increase life expectancy, complications related to both HIV infection and ARV are becoming increasingly prevalent and are gaining importance for those living with HIV infection.

Since early in the epidemic, hypogonadism has been frequently described in HIV positive men. In this chapter, we discuss the available literature in males concerning the prevalence of hypogonadism in HIV disease, the likely etiologies and associations, and discuss the complications and treatment considerations in this group.

G. Cambareri, MD
Division of Urology, Department of Surgery, UMDNJ New Jersey Medical School, Newark, NJ, USA

J. Ashby, MBBS MRCP
Department of HIV and Sexual Health, Berkshire Healthcare NHS Foundation Trust, Slough, UK

D. Goldmeier, MD, FRCP
Jane Wadsworth Clinic, Imperial College NHS, London, UK

H. Sadeghi-Nejad, MD, FACS (✉)
Division of Urology, Department of Surgery, UMDNJ New Jersey Medical School, Newark, NJ, USA

Hackensack University Medical Center, 20 Prospect Avenue, Suite 611, Hackensack, NJ 07601, USA
e-mail: hossein@ix.netcom.com

W.J.G. Hellstrom (ed.), *Androgen Deficiency and Testosterone Replacement: Current Controversies and Strategies*, Current Clinical Urology,
DOI 10.1007/978-1-62703-179-0_9, © Springer Science+Business Media, LLC 2013

Methods

A review was undertaken using a PubMED search using the umbrella terms *HIV* or *AIDS* and *testosterone* or *androgens*, from 1985 onwards. Significant papers found in references in the primary search were also included. A review was initially undertaken of the clinical aspects of naturally occurring androgens in HIV disease androgen therapy was then considered and presented.

Prevalence of Hypogonadism in HIV Positive Males

Pre-Highly Active AntiretroViral Therapy

Early on in the HIV epidemic, in the pre-highly active anti retro-viral therapy (HAART) era, high rates of hypogonadism were reported. A paper from Dobs et al. in 1988 reported that 20/40 (50%) males with AIDS were hypogonadal as assessed by total testosterone levels, of which 75% had hypogonadotropic hypogonadism. Early correlation with lymphocyte depletion and weight loss was noted [4].

In 1991, Raffi et al. observed lower rates of hypogonadism within a cohort of men living with HIV [5]. Ten out of 67 (15%) men had a total testosterone (TT) level <300 ng/mL. Amongst those with an AIDS diagnosis, prevalence of hypogonadism was significantly higher with 8 out of 28 (29%) individuals biochemically hypogonadal. There was significantly more hypogonadism noted in patients with increasing immunocompromise and with an AIDS diagnosis compared to asymptomatic patients and those with other early stage HIV disease. Low testosterone levels were again mainly seen in association with normal or low pituitary hormone levels. Testosterone responses to gonadotropin releasing hormone were investigated and were found to be normal, suggestive of a functional deficit in the hypothalamic–pituitary axis, as opposed to primary gonadal failure. Normal thyroid stimulating hormone (TSH) and adrenocorticotropic hormone (ACTH) levels were observed [5].

SHBG

Early studies of testosterone levels in HIV positive populations used total testosterone levels as a measurement. In 1995, Laudat et al. measured androgen levels along with sex hormone binding globulin (SHBG) levels in 58 asymptomatic HIV positive men (cases) and compared to 11 HIV negative men as controls [6]. SHBG levels were found to be significantly higher as compared to controls, even in early asymptomatic HIV infection. Non-SHBG bound testosterone was measured and found to be lower in HIV cases compared to controls. CD4 expressing T lymphocyte

Table 9.1 Usefulness of TT, FT (RIA) in the diagnosis of hypogonadism in HIV-infected male

	Sensitivity (%)	Specificity (%)	NPV (%)	PPV (%)	R(+)	R(−)	Diag. OR
TT	25.0	100	89.6	100	Infinite	0.75	Infinite
FT (RIA)	33.3	96.2	90.4	57.1	8.76	0.69	12.7

Calculated free testosterone (CFT) <0.22 nmol/L (6.36 ng/dL) used as the reference for diagnosis of hypogonadism

NPV negative predictive value; *PPV* positive predictive value; *R(+)* positive likelihood ratio; *R(−)* negative likelihood ratio; *Diag. OR* diagnostic likelihood ratio; *TT* diagnosis of hypogonadism as total testosterone <10.4 nmol/L (3 ng/mL); *FT (RIA)* diagnosis of hypogonadism as measured free testosterone <0.031 nmol/L (9 pg/mL)

cell counts are used as a proxy of immune function in HIV positive individuals, and a strong reverse correlation with cortisol/DHEA ratios was noted. The authors suggested that this could reflect alterations in adrenal steroid secretion, with a shift from androgen to glucocorticoid production as HIV disease progresses.

The finding of high levels of SHBG was borne out of several subsequent studies. It has been observed that the high concentration of SHBG in this population may frequently result in an increase in total testosterone (TT) values, alongside a reduction in free or bioavailable testosterone. In a recent study, by Moreno-Perez et al., levels of SHBG were compared between 36 HIV positive men aged 40 years or less and a reference population of 127 healthy eugonadal men aged 18–30 years [7]. The median SHBG concentration was significantly higher in the HIV-infected patients (39 nmol/L) compared to the healthy control group (27 nmol/L). In the same study, SHBG, albumin, TT, and free testosterone (FT) were measured. Bioavailable testosterone was calculated using the equation described by Vermeulen et al. [8]. Ninety HIV positive men were included in this arm of the study, comprising 84% who were on ARV treatment and 72% with a suppressed HIV viral load. Using the calculated bioavailable testosterone level as a measure, hypogonadism was observed in just 13% of the population studied. TT and FT were found to have a sensitivity of just 25% and 33%, respectively, in predicting hypogonadism in this population (Table 9.1).

The results of these studies collectively suggest that measurement and calculation of bioavailable testosterone in HIV positive persons is a more accurate way of detecting biochemical hypogonadism, particularly in those with borderline low free or total testosterone levels.

In recent years, since the introduction of effective treatment for HIV in the form of HAART, studies have sought associations between ARV treatment and hypogonadism in HIV positive males.

HAART Era

The effect of ARV therapy on testosterone levels is unclear. Studies looking at the effect of ARV on androgen levels have shown conflicting results.

A study performed in 2007, Crum-Cianflone et al. aimed to establish the prevalence and risk factors for hypogonadism amongst a modern cohort of HIV-infected men [9]. Three hundred HIV positive men were enrolled in this study, 60% of which were on ARV with generally good immune function (mean CD4 cell count of 522 cells/mm^3). All had mean age of 39 and included a mixture of ethnicities. Seventeen percent (50/296) of the men were hypogonadal (as defined by a morning total testosterone level <300 ng/dL), and a further 16% had a borderline testosterone level (300–400 ng/dL) [9]. In multivariate analysis, increasing age and a higher body mass index (BMI) were positively associated with hypogonadism, while smoking was negatively associated (OR 0.44, $p=0.02$). A subset of participants with low testosterone had luteinizing hormone (LH) and FSH measured, all of which were found to be low, suggesting secondary hypogonadism. There was no association observed between hypogonadism and current, past, or cumulative use of HIV medications.

Wunder et al. demonstrated a high prevalence of low free testosterone levels amongst an untreated HIV positive cohort and studied the effect of ARV therapy on androgen levels over time [10]. Data were derived from stored serum samples from the Swiss HIV Cohort Study (SHCS), which is an ongoing multicenter research project. Ninety-seven participants had serum levels of LH, FSH, and FT measured at baseline, and again, after 2 years of successful HAART. At baseline, 68 patients (70%) had subnormal FT levels based on an age-adjusted normal range. Of these, LH levels were low in 44%, normal in 47%, and high in 9%, indicating a mainly secondary hypogonadism in this group. There was a trend for association between lower CD4+ T-cell counts and hypogonadism, but no other associations were found for age, BMI, sexuality, duration of HIV disease, or plasma HIV viral load. No significant changes in free testosterone, LH, or FSH were observed after 2 years of successful ARV therapy. Overall, >60% remained hypogonadal at this point. Twenty-four percent of those who were originally hypogonadal returned to normal levels, and 32% of those originally within normal levels became hypogonadal after 2 years of therapy. All participants on the ARV regime were on the nucleoside reverse transcriptase inhibitors (NRTI) zidovudine and lamivudine. Sixty-three percent were also on a boosted protease inhibitor (bPI) as the third agent, and the remaining 37% were on a non-nucleoside reverse transcriptase inhibitor (NNRTI). The authors acknowledged that the use of stored serum samples was a limiting factor in the interpretation of these findings, as the samples were not consistently morning levels; therefore, this may have led to higher rates of reported hypogonadism.

Effective treatment with ARV has been shown to increase lean body mass (LBM), particularly in individuals with low CD4 counts at initiation of treatment [11]. In a study looking at the effect of some older ARVs on LBM and testosterone levels, Dube et al. noted increases in free testosterone after initiation of ARV, along with an increase in fat-free mass [12]. The rise appeared to be more marked with certain types and classes of ARV, with greater increases at 64 weeks among zidovudine–lamivudine recipients than among stavudine–didanosine recipients, and among efavirenz (an NNRTI) recipients than among nelfinavir (a bPI) recipients [12]. It is worth noting that in recent years, usage of stavudine, nelfinavir, and to a lesser degree didanosine has declined in parts of the world with access to newer ARV agents, due to the high levels of toxicities observed, including lipodystrophy.

In the Massachusetts Male Aging Study, rates of hypogonadism amongst HIV negative men were reported as between 6 and 12% amongst 40–69-year-old American males [13]. Reported prevalence of hypogonadism in HIV positive men varies widely, and rates reported depend on the type of testosterone measurement used; however, it seems clear that prevalence of hypogonadism is higher in HIV positive compared to HIV negative males.

While studies show there are high levels of biochemical hypogonadism in clinical practice, a diagnosis of testosterone deficiency is usually made in conjunction with consistent clinical symptoms and signs [14]. Testosterone deficiency can cause nonspecific symptoms such as fatigue, depressed mood, weakness, and sexual dysfunction, all of which are also frequently associated with many chronic illnesses, HIV virus, ARV therapy, and opportunistic infections. Careful history, examination, and appropriate investigation are essential when trying to ascertain whether biochemical hypogonadism is clinically relevant in an individual.

HIV Wasting and Testosterone

The US Centers for Disease Control and Prevention (CDC) classification system for HIV disease staging and classification defines HIV wasting syndrome as involuntary weight loss >10% of baseline body weight, in association with either chronic diarrhea or chronic weakness and documented fever for ≥1 month. Wasting is further identified as an indicator for the AIDS condition [15]. In the developed world, wasting amongst HIV positive persons taking ARV is now less common and frequently reverses quickly with appropriate HIV viral control and nutritional supplementation. However, even in the era of HAART, HIV-associated weight loss remains prevalent. A large retrospective observational study of a contemporary managed care population in the United States observed that even amongst a modern cohort of HIV, almost one in ten had evidence of HIV-associated weight loss [16].

In a study presented in 2000, by Desyatnik et al., the association between clinical wasting and hypogonadism was investigated [17]. A retrospective review of 88 male patients with HIV infection who had been on HAART for at least 6 months was performed. Hypogonadism was defined as serum-free testosterone <15 pg/mL and was observed in 20% of participants. Wasting was defined as weight loss >5% of ideal body weight and was observed in 42% of the subjects. The investigators observed that testosterone deficiency in these males did not correlate with wasting.

The same year a paper was published by Rietschel et al. looking specifically at androgen levels in a group of HIV positive men with wasting [18]. Seventy percent of participants were receiving ARV therapy. The authors found that in this population with wasting, 21% of patients receiving HAART had hypogonadism compared with 15% of those not on HAART. The prevalence of hypogonadism was not related to any particular class of ARV. Comparison was also made with a control group of

HIV positive men without wasting, and total and free testosterone levels were not significantly different for subjects with wasting and subjects without wasting.

In the context of the developing world, wasting syndromes are commonly encountered and may have multiple etiologies including lack of appropriate ARV, concomitant infections, such as parasitic infection and tuberculosis, and lack of nutritional intake and food security. In this instance, the appropriate remedy of the any exacerbating underlying cause of wasting would be indicated in the first place before consideration of testosterone replacement.

Testosterone, Cytokines, and Wasting

The contributions of catabolic cytokines to HIV-associated wasting and weight loss, in a group of HIV positive men, were investigated by Roubenoff et al. [19]. The authors studied a group of 172 men, 43% of which were on HAART. They found that both TNF-α and IL-1β production by peripheral blood mononuclear cells predicted loss of LBM, and that serum-free testosterone was inversely associated with TNF-α production. However, serum-free testosterone alone was not an independent predictor of either LBM or resting energy expenditure after adjustment for cytokine production. The authors suggested that the inverse relationship between testosterone and cytokine production might be due to testosterone acting as a modulator of the effect of TNF-alpha and IL-1beta and not as a direct determinant of body composition or metabolic rate.

Prevalence of Hypogonadism in Any Chronic Disease

Hypogonadism has been observed in the presence of many common acute and chronic illnesses in men such as chronic liver and renal disease, cancer, and sepsis. In the case of depressed testosterone in HIV disease, the symptoms and signs observed are often nonspecific and may be attributed to the underlying disease or treatment. Low levels of testosterone may cause loss of LBM, bone mass density, and produce symptoms such as depressed mood, loss of energy, and sexual dysfunction. The mechanisms explaining hypogonadism and various systemic diseases are incompletely understood, but these conditions are likely caused by a combination of stress, nonspecific weight loss, inflammation, and medication [20].

Metabolic Syndrome

The metabolic syndrome (MS) is a term used to describe the clustering of risk factors for cardiovascular disease, including abnormal lipids, hypertension, insulin resistance, and intra-abdominal obesity. In HIV positive cohorts, diagnosed prevalence has been shown to vary according to the diagnostic criteria used [21] and has been reported in between 7 and 45% of individuals [22].

The reasons why high rates of metabolic syndrome are seen in HIV positive individuals are not fully elucidated, but are thought to be related to both treatment and disease factors, and several mechanisms have been suggested. HIV protease inhibitor mediated blockade of glucose transport and NRTI mediated mitochondrial toxicity have been well characterized. Additional cellular effects, including the induction of endoplasmic reticulum and oxidative stress, altered adipocytokine secretion. Lipotoxicity has been implicated in the development of metabolic syndrome in this cohort [23]. In HIV negative cohorts, metabolic syndrome is associated with low levels of testosterone [24, 25].

A recent study by Monroe et al. compared HIV-uninfected men with HIV-infected men and found that had they had lower FT, higher SHBG, and more insulin resistance and DM [26]. Lower FT and lower SHBG were associated with insulin resistance regardless of HIV serostatus. The authors suggested that sex hormones play a role in the pathogenesis of glucose abnormalities among HIV-infected men.

Lipodystrophy

Lipodystrophy syndrome has been reported in as many as 41% of people with HIV and comprises changes in body fat distribution including both fat loss (lipoatrophy) and fat accumulation (lipohypertrophy) [27]. Mitochondrial toxicity and dysfunction have been implicated as a cause for the adverse metabolic toxicities of ARV treatments, including lipodystrophy and insulin resistance. Toxicity caused by the ARV is drug class specific and in the case of NRTI have been attributed to the drugs, causing an inhibitory effect on DNA polymerase gamma enzyme [28].

Sex hormones including testosterone may play a role in altered fat distribution and insulin sensitivity of male patients with HIV lipodystrophy.

Gynecomastia and Hypogonadism

A study by Biglia et al. found that 40 out of 2,275 (1.8%) men with HIV have gynecomastia. Mean free testosterone index (ratio of total *testosterone* divided by SHBG) was significantly lower in those patients with gynecomastia compared to a control group without gynecomastia. The authors conclude that gynecomastia in this group is associated with hypogonadism. Duration of exposure to ARV was not associated with gynecomastia [29]. The finding of lower free testosterone index and bioavailable testosterone levels in patients with gynecomastia has been observed in subsequent studies and the effect of certain ARV treatments. In particular, the NNRTI, efavirenz, has also been associated with gynecomastia in HIV positive men [30, 31].

Abnormal Testosterone Metabolism and Raised Estradiol

There is evidence that men on ARV develop low sexual desire that can be associated with raised estradiol levels. It has been suggested that abnormal metabolism seen in HIV positive men results in increased aromatization of testosterone to estradiol [32]. It has been postulated that increased conversion of testosterone to estradiol may result from increased central adipose tissue seen in lipodystrophy/metabolic syndrome. It is one possible explanation why metabolic syndrome/lipodystrophy syndrome and low testosterone frequently coexist in HIV positive men.

Sexual Function and HIV

High rates of sexual dysfunction have been well described amongst men with HIV. In particular, erectile dysfunction (ED) is prevalent, and has been reported to be present in between 9 and 74% of HIV positive men [33]. Guaraldi et al. studied 133 HIV positive men and found a prevalence rate of erectile dysfunction in 55 and 65% of men aged less than or more than 50 years old, respectively. Comparison of total and free testosterone was made in those with and those without ED, and no significant differences were observed.

In the large study by Crum-Cianflone, ED was reported in 175/285 (61%) [9]. In this study, rates of hypogonadism were not significantly different between those with and those without ED. Increasing age and depression were associated with ED and high CD4 cell count was observed to be protective. There was also no association between either ED or hypogonadism and the current, past, or cumulative use of HIV medications. Similarly, in a cross-sectional study by Lallemand et al., 111 out of 156 (71%) HIV positive men on ARV reported sexual dysfunction. There was also no association between sexual dysfunction and types of ARV used [34].

The study by Moreno-Perez et al. revealed that 100% (12/12) of patients with hypogonadism, as measured by calculated free testosterone (CFT), had ED, but only a quarter of all patients with ED were hypogonadal [7].

Using CFT or bioavailable testosterone levels to diagnose hypogonadism is an accurate way of identifying ED related to or worsened by low testosterone levels. However, the cause of ED is often multifactorial; thus, the use of biochemical hypogonadism to investigate ED is a valuable part of the diagnostic armory available when investigating HIV positive patients with ED.

Age

Hypogonadism is associated with increasing age. Total testosterone decreases year-on-year in HIV negative populations as they move through their fourth and fifth decades of life [13, 35]. Within developed countries, we are now caring for an aging cohort of HIV positive individuals, and it is predicted that as prognosis continues to improve for individuals living with HIV, there will be increasing numbers of older people living with HIV [36, 37]. We may reasonably expect that the numbers of HIV positive men with late onset hypogonadism increase in the future. In 2005, Klein et al. set out to examine the prevalence and association of hypogonadism in males at risk of or living with HIV aged over 50 years old [38]. In a sample of over 500 men, aged >49 years, of which 275 were HIV positive, the authors noted that hypogonadism was prevalent. The HIV positive group was associated with high HIV viral load, but not with CD4 count or disease stage. Raised total testosterone was associated with HIV positivity, but free androgen index (FAI) was not associated with HIV status. Both low, free androgen index and low androgen levels were correlated with clinical findings suggestive of hypogonadism.

There is a growing body of literature describing a process of accelerated aging in HIV-infected individuals. Chronic immune activation and inflammation are observed, along with thymic dysfunction and gut microbial translocation. All act together to drive early senescence in HIV infection [39]. Among the HIV positive women, early loss of ovarian function and early menopause have been described in some study populations [40–42] as well as an unexpectedly high numbers of HIV positive women reaching menopause very early [40, 43]. It may be that men living with HIV infection are more prone to a syndrome of late onset androgen deficiency or an "early andropause," heralded by dysregulation of the hypothalamic–pituitary axis as part of a process of accelerated physiological aging [44].

Bone Mineral Density and Testosterone

High rates of osteopenia and osteoporosis have been reported in HIV-infected cohorts. Prevalence rates range between 22–71% and 3–33%, respectively [45]. Loss of bone mineral density (BMD), in the context of HIV, is likely to be multifactorial, and contributing factors include ARV treatments (particularly implicated is the widely used NRTI, tenofovir), vitamin D deficiency, alcohol and drug use, smoking, low BMI, and HIV infection. Hypogonadism may be an additive risk factor for low BMD and should be managed appropriately, because part of the osteoporosis management includes measures to improve calcium and vitamin D nutrition, increase weight-bearing exercise, and specific pharmacologic therapy, including, in some cases, testosterone replacement [46].

Recreational Drug Use

Intravenous drug use is a risk factor for HIV acquisition, and recreational drug use is not uncommon among HIV-infected cohorts. Marijuana, opiates, anabolic steroids, and alcohol can all inhibit gonadal function [47]. A study on androgen levels, by Wisniewski, showed that free T concentrations were lower in men who used cocaine and/or opiates, irrespective of HIV status [48].

Treatment of Hypogonadism and Wasting Syndrome in HIV Positive Men

There are various treatments available for hypogonadism and wasting syndrome in HIV positive men. The literature is wide ranging on the multitude of therapy options and benefits seen in HIV men. Treatment of hypogonadism and wasting syndrome in HIV positive men has been shown to help with LBM, weight loss, depression, fatigue, libido, and BMD.

Therapy ranges from testosterone replacement to synthetic analogues such as anabolic steroids as well as growth hormone, dronabinol, megestrol acetate, and insulin-like growth factor I. In this section, treatment for hypogonadism and wasting syndrome in HIV positive men will be reviewed. The review will include various types of treatments, methods of delivery, benefits, and side effects.

Body Composition and Muscle Wasting

Body composition may undergo drastic changes in HIV positive patients with losses in LBM, muscle mass, and associated wasting syndrome. The latter is defined by the CDC as the involuntary loss of more than 10% of baseline body weight in the previous 12 months [49, 50]. Body cell mass (a component of LBM) [50] represents metabolically active tissue, and men with HIV wasting may show losses in body cell mass and LBM prior to overall changes in body weight and fat mass [51–53]. These measured changes are a better indicator of wasting syndrome and malnutrition in men with HIV. Losses of LBM and muscle mass have been shown to correlate with androgen levels in hypogonadal men with HIV, and research has focused on different treatment options to counteract this loss [54].

Androgen replacement therapy in this population has been extensively studied and includes different forms of synthetic testosterone, in addition to metabolic steroids. The literature has focused on men with HIV with and without wasting, HIV positive eugonadal men and hypogonadal men. In addition to androgen replacement, other treatment options that have been studied in relation to body composition include megestrol acetate, dronabinol, insulin-like growth factor I, and growth hormone. Resistance training and exercise have also been studied along with testosterone replacement as well as whether there are additive effects when used in combination.

Types of Therapy

Megestrol Acetate

Megestrol acetate has been shown to benefit men with wasting syndrome. Men randomized to a 12-week randomized controlled study taking either placebo, 100, 400, or 800 mg of megestrol acetate were shown to have significant increases in total weight at 800 mg of megestrol acetate compared to placebo, with overall trends toward increasing weight in a dose-dependent fashion [55]. The total weight gain average in the 800 mg group was 3.54 kg. This was compared to a weight loss of 0.725 kg in the placebo group ($p<0.001$), and average LBM gain of 1.14 kg in the 800 mg group was compared to 0.772 kg weight loss in the placebo group ($p<0.001$). A similar study, which compared dosing of 800 mg of megestrol acetate to placebo, found significant total weight gain in the therapy group compared to placebo; however, this was due to gains in fat content as opposed to LBM, which was not significant between the groups [56]. A recent study compared megestrol acetate therapy to oxandrolone therapy and found similar weight gains between the groups (2.8 kg and 2.5 kg, respectively, $p=0.80$) and LBM (39% and 56% of weight, respectively, $p=0.38$). Adverse events between the groups were not significant ($p=0.74$) [57].

Dronabinol

Dronabinol is an orally active cannabinoid with complex central nervous system effects that increased the appetite of patients with HIV wasting. In a study of 139 patients, randomized to dronabinol or placebo, weight remained stable in the dronabinol group compared to an average weight loss in the placebo group of 0.4 kg. However, increased appetite above baseline and decreased nausea were observed compared to placebo [58]. In an open-label study designed to assess safety and pharmacokinetics of dronabinol and megestrol acetate, alone or in combination in HIV wasting syndrome, dronabinol was found to have no effect on weight (mean weight change −2.0 kg) compared to megestrol acetate alone (6.5 kg) [59].

A more recent retrospective review of 117 patients treated with dronabinol over the course of 3 months to 1 year showed that mean weight gain in patients taking the medicine for 1 year was 3.7 lb. Loss of appetite was also improved as well as nausea [60]. Perhaps, more long-term prospective studies can assess the effect of dronabinol on weight gain and changes in body composition such as LBM.

Growth Hormone

Growth hormone (GH) has been shown to change body composition in HIV positive men. LBM and total weight are increased, although the gains are not often sustained [61]. GH has been approved by the FDA to treat wasting associated with HIV virus, but the cost can be prohibitive [50]. One hundred and seventy-eight men randomized to receive either growth hormone (0.1 mg/kg) or placebo showed significant gains in LBM (3.0 kg, $p=0.001$), total weight (1.6 kg, $p=0.001$), and treadmill work output in the treatment group over the course of 12 weeks. Additionally, body fat differences were observed between the groups with a significant decrease in the treated patients (−1.7 kg, $p<0.001$) [62]. Side effect differences between the two groups were significant and included swelling or puffiness and arthralgias in the treatment arm. Additionally, greater increases in blood glucose and HbA1c were observed in the GH group.

Further studies, including usage of lower doses of growth hormone (30–40 μg/kg), have found similar observations with increased LBM, decreased fat mass, and decreased trunk and appendicular fat mass [63, 64]. Significant side effects were seen, as in previous studies, including acute increases in fasting glucose levels [62]. An open-label study, which administered growth hormone at a dose of 6 mg/day for 24 weeks followed by a 12 week washout period and then 4 mg dosing every other day for 24 weeks, found significant decreases in visceral adipose tissue, although the effect was not great for the lower dose (visceral adipose tissue decreased an average of 42% at 12 weeks during the high dose treatment vs. 15% in those who continued another 12 weeks at the lower dose). Forty-three percent ($n = 13$) of participants experienced side effects during the trial, and 11 patients in the high dose phase and 3 in the low dose either temporarily or permanently discontinued treatment [65].

Side effects of growth hormone were also highlighted in a comparison study of patients randomized to receive either nandrolone decanoate (150 mg IM biweekly), a metabolic steroid, or placebo in a double-blind study, which was compared to an open-label group of patients receiving growth hormone (6 mg daily). Nandrolone and growth hormone were associated with greater increases in LBM compared to placebo (1.6 and 2.5 kg, not significant) as well as greater gains in fat-free mass. Growth hormone also showed greater decreases in whole body fat mass, as seen in previous studies; however, a greater percentage of patients (47.6%) experienced adverse drug events compared to the nandrolone (4.7%) and placebo (4.8%) groups. Most frequently reported events included edema, arthralgia, and carpal tunnel syndrome [66].

In perhaps the largest study to date, including 757 subjects, a randomized double-blind placebo-controlled trial compared patients who received a maximum of 6 mg of growth hormone daily (DD) or as alternate day (AD) dosing. Body weight was shown to increase by 2.9 kg and 2.2 kg, respectively. The median increase in LBM after 12 weeks was 3.3 kg in the AD group and 5.2 kg in the DD group (for both groups: $p < 0.0001$ vs. placebo). Greater decreases were also seen in total fat mass and truncal fat mass compared to placebo, with greater decreases in the daily dosing compared to the alternate dosing regimen [67].

In a meta-analysis of randomized placebo-controlled studies, growth hormone was found to improve LBM by approximately 3 kg compared with placebo, and was associated with improvements in physical endurance. However, as previously reported, adverse events were more common including arthralgias, edema, myalgias, and increases in blood glucose [68].

Insulin-Like Growth Factor I

Insulin-like growth factor I is not effective alone in the treatment of wasting associated with HIV, and is associated with detrimental metabolic effects including hypoglycemia [69, 70]. In combination with growth hormone, decreases in fat-free mass and body fat have been noted, although no significant changes in LBM were observed [71]. Another study documented early increases in weight that were not observed at later time points, as well as transient increases in fat-free mass without an overall significant anabolic effect [72]. LBM was observed to increase with combination therapy in another study which randomized patients to four groups: growth hormone or insulin-like growth hormone I alone, in combination and placebo. The combination treatment group showed the greatest gain in LBM, 3.2 kg ($p<0.001$); this was the only group that showed changes sustained at 12 weeks [69].

Testosterone

Hypogonadism is prevalent in HIV positive men, as high as 50% in some studies, including 25% with primary gonadal failure (although with the advent of HAART, the percentages may be lower) [4, 54]. Additionally, there is a correlation between exercise capacity and testosterone levels that highlight the important anabolic effects of testosterone, and support a potential role in wasting [54]. Although not meeting the technical center for hypogonadism patients experience an acute drop in testosterone levels, when in the normal range may experience symptoms [73]. Testosterone replacement therapy in HIV positive men has been well studied, although many of them have small patient populations and are not uniform in outcome measures. Eugonadal and hypogonadal men as well as men with and without wasting syndrome have been studied for testosterone replacement.

Testosterone has been shown to influence body composition and change LBM, which is the metabolically active component of body composition. The use of testosterone in wasting syndrome may be equally efficacious as in men in whom wasting has not occurred, with the added benefit of anabolic effects. Different types of testosterone have been studied including intramuscular formulations, transdermal/transscrotal patches, and testosterone gel.

Intramuscular Testosterone

Although testosterone administration has been shown to have no benefit on weight in one study [74], the majority of studies have shown increases in weight and overall changes in body composition. In a double-blind placebo-controlled trial of 51 HIV positive men with evidence of wasting and hypogonadism, patients were randomly assigned to receive testosterone enanthate 300 mg, every 3 weeks or placebo. Gains in fat-free mass were seen in the placebo vs. treatment group at 6 months (−0.6 and 2.0 kg, $p=0.36$), as well as LBM (0 and 1.9 kg, $p=0.041$) and muscle mass (−0.8 and 2.4 kg, $p=0.005$). Testosterone was tolerated well by patients, and significant benefits were also seen in perceived improved quality of life, improved appearance, and feeling better [75].

The above patient population was then followed during an open-label treatment trial where the placebo group crossed over to receive testosterone [76]. Only after crossover did the placebo group show gains in LBM (–0.6 kg at 0–6 months compared to 1.9 kg at 6–12 months; $p=0.03$). Subjects who initially received testosterone and continued in the study continued to gain LBM with greater gains at 1 year compared to subjects who initially received placebo and then crossed over (0–6 months 1.6 kg gained compared to 3.7 kg at 6–12 months; $p<0.05$).

In another study, which looked at testosterone replacement in HIV men with and without hypogonadism (testosterone levels at the lower end of normal), primary end points included low libido, mood, energy, and weight [77]. Men were required to have low sexual desire as inclusion criterion for entry into the study. An open-label design was used with testosterone cypionate 200 mg initially and 400 mg biweekly for 8 weeks of open therapy, followed by an additional 4 weeks of treatment for the responders. They then underwent randomization to a placebo-controlled, double-blind treatment arm. Average weight gain was 3.5 lb ($p<0.001$); however, body composition studies were not performed. Significant improvements were also seen with mood and sexual desire/libido. Of the men in the treatment arm randomized to placebo, only 13% of those in the placebo group maintained their response as compared to 78% in the treatment group. Importantly, men who had low (but within the reference range) levels of testosterone were just as likely to respond as hypogonadal men.

A similar study of men with and without hypogonadism, with inclusion criteria requiring sexual dysfunction (weight loss was not a requirement), was performed in 70 men in a double-blind randomized placebo-controlled trial. Men were assigned to a placebo treatment arm or injections of testosterone cypionate 200 mg initially and 400 mg biweekly. At 12 weeks, the men were assigned to open-label maintenance. Along with improved mood, energy, and libido at the 12-week mark, average weight gain was 2.6 kg ($p<0.01$) in the treatment group with gains in muscle mass of 1.6 kg ($p<0.001$). The increase was greater for men with initial wasting at presentation (2.2 kg, $p<0.001$) [73].

High doses of testosterone enanthate (300 mg/week) were utilized in a randomized, double-blind, placebo-controlled trial of HIV positive men with weight loss. Significant increases in fat-free mass were seen in the testosterone group compared to the placebo group (2.8 kg, $p<0.0001$). Interestingly, serum testosterone levels correlated with the increase. Total body weight also increased within the treatment arm (1.8 kg, $p=0.003$), although it was not significantly different compared to placebo [78].

Two studies compared men in a 2×2 factorial design, combining resistance training with or without testosterone treatment and placebo with or without testosterone treatment to determine if there were additive benefits to changes in body composition [79, 80]. The first study included hypogonadal men with weight loss. Body weight was noted to increase significantly in the testosterone group (testosterone enanthate, 100 mg/week) (2.6 kg, $p<0.001$) and in the exercise group alone (2.2 kg, $p=0.02$), however, not in the placebo group or combined testosterone group.

Maximum strength in various exercises including leg press, leg curls, bench press, and latissimus pulls was found to increase in the testosterone-exercise group, but not in the placebo group. LBM increased by 2.3 kg ($p=0.004$) and 2.6 kg ($p<0.001$), respectively, in the testosterone alone or testosterone-exercise group but not in the placebo group. Testosterone and exercise combined did not produce greater results than either one alone [79].

In a similar trial utilizing testosterone enanthate 200 mg/week in eugonadal men with evidence of wasting [80], men treated with testosterone and testosterone-resistance training were found to have greater increases in LBM compared to placebo or placebo-resistance training alone (4.2 kg, 4.6 kg, 0 kg and 2.3 kg, respectively). Significant increases in LBM ($p=0.05$) and muscle area ($p<0.05$) were observed in the resistance training group alone, gains similar to those seen in low dose testosterone replacement. Similar increases in weight, decreases in fat mass and arm and leg muscle area were observed across groups.

In a meta-analysis of testosterone therapy (given in different forms, i.e., intramuscular, transdermal) in HIV wasting syndrome, eight trials met inclusion criteria for a total of 417 patients. A difference in LBM was seen in the testosterone group compared to the placebo group (1.22 kg, CI 0.23–2.22); however, the difference was greater in trials which utilized intramuscular testosterone (3.34 kg). Total body weight increased by 1.04 kg compared to the placebo group. When women were excluded from the trials, the overall increase in LBM, fat-free mass, or body cell mass was 1.99 kg ($p=0.03$). Excluding women from the analysis, total body weight increased by 1.54 kg ($p=0.05$) [81].

Overall, these studies show that intramuscular injection of testosterone is effective in changing body composition in hypogonadal and eugonadal men with wasting. Although total weight gain is an important element in treating men with HIV-associated wasting, changes in body composition are perhaps more important because it may represent increases in metabolically active tissue as opposed to gains in fat mass. Additionally, the effects on body composition are sustained in men on testosterone replacement therapy who continue long-term treatment. Shortcomings in many of these trials are lack of standardization and outcome measures making meta-analysis difficult, especially when evaluating changes in quality of life parameters and mood, objective changes in exercise/training, and sexual desire and libido [81].

Testosterone undecanoate (TU) is the first long-acting injectable testosterone formulation that has shown to maintain stable testosterone levels in hypogonadal men with an excellent safety profile [82]. This testosterone is available in Europe, but has not been approved for use in the United States at the time of this writing. A review of the database at the national library of medicine does not show any publications on the effects of long-acting intramuscular testosterone formulations (i.e., testosterone undecanoate—TU) on HIV-related hypogonadism.

Transscrotal Testosterone

Only one study to date has looked at the effects of a transscrotal delivery system of testosterone in HIV men with wasting and hypogonadism. One hundred and thirty-three men were randomized in a double-blind placebo-controlled trial to receive a 15 mg transscrotal testosterone patch (designed to release 6 mg of testosterone daily) or placebo patches. Although increases in morning serum testosterone and free testosterone were noted in the treatment arm, no changes in body cell mass or total weight were seen between the groups. The authors speculate that the delivery system did not sustain peak levels of testosterone or maintain levels of testosterone throughout the day. There were also no changes in quality of life [83].

Transdermal Testosterone

Transdermal testosterone has been studied in women with minimal effect [84]. In a study of Androderm (a nongenital transdermal system), in men with HIV and testosterone levels less than 400 ng/dL (men were not required to have evidence of wasting), men were assigned to receive two patches for a total of 5.0 mg/patch over 24 h or placebo. Although serum total and free testosterone levels increased in the treatment arm, there were no significant differences in LBM or fat-free mass between the groups; however, within the treatment arm increases in LBM and fat-free mass were significant (1.34 kg, $p=0.02$ and 1.36 kg, $p=0.02$, respectively). The testosterone group did show significant decreases in fat mass compared to the placebo group ($p=0.04$) and there were no changes between the arms in overall quality of life [85].

Testosterone Gel

Two studies have evaluated testosterone gel in men with HIV with one study comparing the gel formulation to intramuscular testosterone injections. Bhasin et al. [86] looked specifically at HIV hypogonadal men with abdominal obesity. Men were randomized to receive 10 g testosterone gel daily (AndroGel) or placebo for 24 weeks. Although visceral fat did not show a significant difference between the groups, total and subcutaneous fat mass decreased in testosterone treated men compared to placebo ($p=0.04$ and $p<0.001$, respectively). Additional significant decreases were seen in whole body, trunk, and appendicular fat mass ($p<0.001$). Greater increases were also seen in LBM (1.3 kg treatment group vs. –0.3 kg placebo group, $p=0.02$).

One argument for testosterone gel is ease of use, and there is some evidence that there are less fluctuations in daily serum testosterone levels with gel administration as compared to injectable formulations. In a trial of 30 patients in an open-label study, men were given IM testosterone cypionate 100–200 mg every 1–2 weeks, with dose titrations to achieve eugonadal levels. Men were then switched after 8 weeks to AndroGel 5 g daily with dose titrations depending on clinical response. Total treatment time was 16 weeks. Overall therapeutic levels of testosterone were greater in the gel group, with less fluctuation between the peak and trough levels as compared to IM testosterone. However, no significant changes in LBM were observed [87].

Anabolic Steroids

Oxandrolone

Testosterone analogues, specifically oxandrolone and nandrolone, have been studied extensively in HIV men with wasting. There is evidence that anabolic steroids have greater anabolic effects than testosterone and promote greater increases in muscle and LBM [51]. The most important adverse event of steroids with high anabolic potential is liver toxicity, and many studies report increases in serum liver function tests.

In a study of 63 HIV positive men with evidence of wasting, subjects were randomly assigned to 15 mg of oxandrolone/day, 5 mg of oxandrolone/day, or placebo [88]. Baseline testosterone values were not reported. The authors found the 15 mg/day group had significant increases in weight (1.5 lb) compared to the 5 mg/day group (maintained their weight) and the placebo group (continued to lose weight). Body composition data were not reported. The authors concluded that further studies were needed to assess higher dosing of oxandrolone.

In an ongoing large study of 572 patients, open-label treatment of 20 mg of oxandrolone daily was given to men with HIV-associated weight loss. Body composition studies were determined at 1, 2, 4, 8, and 12 months. Only 26 patients were evaluable at 12 months. Mean increases in body weight ranged from 1.9 kg at 1 month to 5.2 kg at 12 months. There were also significant increases in body cell mass and overall the treatment was well tolerated [89].

In a randomized double-blind placebo-controlled trial, 262 eugonadal HIV men with associated wasting were randomized to placebo, 20, 40, or 80 mg of oxandrolone daily. Men were followed for 12 weeks, and then offered the option of continuing in an open-label study of oxandrolone 20 mg daily for 12 weeks. One hundred and ninety-five men completed the randomized trial, and 193 men completed the open-label trial. At 12 weeks, the only significant gain in body weight was seen in the 40 mg group (2.8 kg, $p=0.004$) compared to placebo. The only significant gain in body cell mass was seen in the 40 and 80 mg group compared to placebo (1.5 kg, $p=0.0049$ and 1.8 kg $p=0.0002$, respectively). Laboratory abnormalities in ALT/AST were seen in a dose-dependent fashion with grade III and IV liver toxicities seen. Three patients in the 40 mg group and four in the 80 mg group were discontinued from the study for this reason [90].

Nandrolone

The metabolic steroid nandrolone has shown promise in men with HIV wasting. In a study of HIV positive men with evidence of wasting and failure to gain weight despite nutritional intervention ($n=24$ out of 220), nandrolone decanoate

(100mg/mL) was administered every 2 weeks with body composition assessments performed. After 16 weeks, men in this study showed significant gains in total weight (0.14 kg/week, $p<0.05$) and LBM (3 kg, $p<0.005$). Additionally, changes in quality of life parameters were significantly changed. No major adverse events were noted [91].

Progressive resistance training has shown to benefit body composition in men on testosterone therapy. To determine if it would also have additive effects for men on nandrolone, a group of 20 HIV positive eugonadal men with no evidence of wasting were randomly assigned to receive nadrolone at 200 mg for week 1, 400 mg week 2 and then 600 mg for weeks 3–12. Increases in total body weight, body cell mass, muscle size, and strength were seen within each group for the nandrolone alone subjects and for the combined nandrolone and resistance-training subjects. The magnitude of increase between the groups was significant for LBM (3.9 kg vs. 5.2 kg, $p=0.03$).

Muscle strength change (based on repetitive exercises) was 10.3–31% in the nandrolone alone group and 14.4–53.0% in the nadrolone-resistance training subjects [92]. Overall, similar to studies which utilize resistance training with testosterone therapy, exercise may have additive benefits on body composition in HIV males. No major adverse events were reported, and liver function tests remained on average at the upper limits of normal.

Oxymetholone

Oxymetholone is a steroid that has high anabolic potency compared to its androgenic effects (8.75:1). In a phase III double-blind, randomized, placebo-controlled study of 89 patients, including men and women, patients were assigned to oxymetholone 50 mg BID, TID, or placebo [93]. The study was conducted for 16 weeks, at which time patients went on to open-label treatment with oxymetholone 50 mg BID for weeks 16–32. Patients had to have 5% or greater weight loss over the preceding 6 months or loss of 10% below ideal body weight in the preceding 12 months. Body composition, side effects, and quality of life measurements were some of the end-points. Total body weight gain was observed in the BID and TID oxymetholone groups (3.0 and 3.5 kg, $p<0.05$ for both compared to placebo). LBM was increased significantly in both groups; however, the BID group gained 2.9 kg compared to the 1.8 kg seen in the TID group. The authors postulate that the BID group had greater gains due to higher numbers of treatment interruptions in the TID group.

Body cellular mass was also significantly increased in the BID and TID groups compared to placebo (3.8 kg and 2.1 kg, $p<0.0001$ and $p<0.005$, respectively), again with greater gains in the BID group probably due to treatment disruptions. Quality of life parameters included appetite and food intake, improved well-being, and decreased weakness and fatigue. All parameters were significantly improved in the treatment groups (no significant changes in the placebo group were observed). Patients who completed the open-label study of oxymetholone 50 mg a day were observed to gain weight if initially in the placebo group and maintained their weight when initially in the treatment group.

Side effects during the randomized double-blind phase included grade III and IV liver toxicities (AST or ALT greater than 5× baseline), which was dose dependent. Liver-related side effects caused treatment disruptions in 16% in the TID group and 3% in the BID group. Other side effects included GI complaints, changes in libido, muscle cramps, acne, and clitoris enlargement in females. Grade III and IV liver toxicity was also observed in the open-label phase in 39% of patients.

Anabolic steroids show promise in treating both hypogonadal and eugonadal men who show evidence of HIV-associated wasting. Whether treatment with these medications is more effective than testosterone replacement remains to be seen. Associated side effects, such as liver toxicity, which is a well-known complication of anabolic steroids, may preclude them as standard treatment in this already immunocompromised population. More long-term studies are needed to fully elucidate long-term effects of anabolic steroids in this study population.

Depression

Depression among those with HIV is prevalent and is multifactorial in origin [94]. The relationship between depression and testosterone levels in hypogonadal men has been studied to further investigate targeted therapy. Major quality of life parameters that have been measured include mood, sexual dysfunction, libido, fatigue, weakness, appetite, food intake, irritability, ability to cope, and performance to name a few. Many of the above studies report quality of life parameters as part of the overall benefit of androgen replacement therapy; however, not all studies have found positive effects [74, 88].

One of the major difficulties in studying quality of life parameters includes standardization across studies. Many indices are used to evaluate quality of life parameters including depression in these subjects. As a result, meta-analysis of these studies is impossible [80]. Additionally, many of these studies do not use quality of life parameters as part of the inclusion criteria or have not selected patients specifically with depression.

The relationship between testosterone levels and depression was evaluated in a randomized placebo-controlled study [95]. Initially depression scores were compared between eugonadal and hypogonadal men. The Beck Depression Inventory was used as the index to evaluate depression among subjects. The authors found that testosterone levels were significantly ($p=0.01$) and inversely associated with the Beck score, with a Beck score >18 (moderately to markedly increased depression) as the cutoff. Subjects were then randomized in a placebo-controlled trial to testosterone enanthate 300 mg IM every 3 weeks as previously described [75]. The Beck score decreased significantly in the treatment arm (−5.8, $p<0.001$) but did not in the placebo-treated patients. The change in Beck score was also highly related to weight changes among the subjects in the treatment arm.

Another study compared fluoxetine and testosterone in HIV men with major depression, subthreshold major depression or dysthymia [96]. Men were randomized to receive either 400 mg IM testosterone cypionate biweekly, fluoxetine 60 mg/day or double placebo. Among the subjects, mood response rates were not statistically significant between the groups. However, improvements in fatigue among the testosterone group were significant (39, 56, and 42% among those who completed the study, $p < 0.05$). The authors conclude that testosterone should not be a first-line treatment in HIV men with associated depression.

Bone Density

Osteopenia in HIV positive subjects is due to both the disease itself and the treatment with antiretroviral medications [97]. The role of testosterone and BMD is well established. Some endocrinologists advocate testosterone treatment in men with evidence of osteopenia [98]. The role in HIV is less well studied, especially among hypogonadal men, although increases in BMD have been shown in HIV negative hypogonadal men [99].

One study looked at testosterone replacement in eugonadal HIV positive men with wasting syndrome [100]. Men were assigned to either testosterone enanthate 200 mg/week or placebo and to progressive resistance training or none in a 2×2 factorial study design. The authors found HIV men had lower baseline lumbar spine and hip BMD. Testosterone replacement significantly increased BMD over 3 months; however, resistance training had no effect. No increases in BMD were seen in the hip or femoral neck region. More studies are needed to analyze the effect of testosterone in HIV positive men, especially in those with concomitant hypogonadism.

Summary Points

- Androgen deficiency is more common in HIV-infected males than in the general population.
- Prevalence of hypogonadism varies widely and depends on the type of measurement of testosterone used.
- Hypogonadism is predominantly secondary to hypothalamic–pituitary axis dysfunction associated with low LH and FSH levels, and not due to primary testicular failure.
- In the era of HAART, rates of hypogonadism may be decreasing compared to early in the HIV epidemic, but men with HIV infection remain at risk due to complications of HIV infection, comorbidities and toxicities associated with HAART, and the aging process.
- Causes of low serum testosterone levels are complex and have been attributed to factors including chronic illness, HIV virus, medications used to treat HIV,

opportunistic infections, comorbidities and coinfections, and normal age-related declines.

- Low levels of testosterone have been variably associated with low CD4 cell counts, high HIV viral loads, advancing age, increasing length of time diagnosed with HIV, disease progression, loss of LBM, metabolic syndrome, wasting, lipodystrophy, and recreational drug use.
- Symptoms and signs of androgen deficiency in HIV positive males are similar to those seen in HIV negative males and overlap with symptoms commonly seen in HIV infection, regardless of androgen status. Symptoms include sexual dysfunction, fatigue, weakness, depression, and accelerated loss of BMD.

References

1. Palella Jr FJ, Delaney KM, Moorman AC, et al. Declining morbidity and mortality among patients with advanced human immunodeficiency virus infection. HIV Outpatient Study Investigators. N Engl J Med. 1998;338:853–60.
2. Detels R, Munoz A, McFarlane G, et al. Effectiveness of potent antiretroviral therapy on time to AIDS and death in men with known HIV infection duration. Multicenter AIDS Cohort Study Investigators. JAMA. 1998;280:1497–503.
3. UNAIDS: World Health Organization. AIDS epidemic update: December 2009. 2009. http://www.unaids.org/en/media/unaids/contentassets/dataimport/pub/report/2009/jc1700_epi_update_2009_en.pdf. Accessed 26 Apr 2012.
4. Dobs AS, Dempsey MA, Ladenson PW, et al. Endocrine disorders in men infected with human immunodeficiency virus. Am J Med. 1988;84:611–6.
5. Raffi F, Brisseau JM, Planchon B, et al. Endocrine function in 98 HIV-infected patients: a prospective study. AIDS. 1991;5:729–33.
6. Laudat A, Blum L, Guechot J, et al. Changes in systemic gonadal and adrenal steroids in asymptomatic human immunodeficiency virus-infected men: relationship with the CD4 cell counts. Eur J Endocrinol. 1995;133:418–24.
7. Moreno-Perez O, Escoin C, Serna-Candel C, et al. The determination of total testosterone and free testosterone (RIA) are not applicable to the evaluation of gonadal function in HIV-infected males. J Sex Med. 2010;7:2873–83.
8. Vermeulen A, Verdonck L, Kaufman JM. A critical evaluation of simple methods for the estimation of free testosterone in serum. J Clin Endocrinol Metab. 1999;84:3666–72.
9. Crum-Cianflone NF, Bavaro M, Hale B, et al. Erectile dysfunction and hypogonadism among men with HIV. AIDS Patient Care STDS. 2007;21:9–19.
10. Wunder DM, Bersinger NA, Fux CA, et al. Hypogonadism in HIV-1-infected men is common and does not resolve during antiretroviral therapy. Antivir Ther. 2007;12:261–5.
11. Shikuma CM, Zackin R, Sattler F, et al. Changes in weight and lean body mass during highly active antiretroviral therapy. Clin Infect Dis. 2004;39:1223–30.
12. Dube MP, Parker RA, Mulligan K, et al. Effects of potent antiretroviral therapy on free testosterone levels and fat-free mass in men in a prospective, randomized trial: A5005s, a substudy of AIDS Clinical Trials Group Study 384. Clin Infect Dis. 2007;45:120–6.
13. Araujo AB, O'Donnell AB, Brambilla DJ, et al. Prevalence and incidence of androgen deficiency in middle-aged and older men: estimates from the Massachusetts Male Aging Study. J Clin Endocrinol Metab. 2004;89:5920–6.
14. Bhasin S, Cunningham GR, Hayes FJ, et al. Testosterone therapy in adult men with androgen deficiency syndromes: an Endocrine Society clinical practice guideline. J Clin Endocrinol Metab. 2006;91:1995–2010.

15. Castro KG, Ward JW, Slutsker L, et al. Revised classification system for HIV infection and expanded surveillance case definition for AIDS among adolescents and adults. 18 Dec 1992. http://www.cdc.gov/mmwr/preview/mmwrhtml/00018871.htm. Accessed 26 Apr 2012.
16. Siddiqui J, Phillips AL, Freedland ES, et al. Prevalence and cost of HIV-associated weight loss in a managed care population. Curr Med Res Opin. 2009;25:1307–17.
17. Desyatnik M, Baaj A, Fisher A. The prevalence of hypogonadism in HIV-infected patients receiving HAART. TuPeB3180. Presented at World AIDS 2000.
18. Rietschel P, Corcoran C, Stanley T, et al. Prevalence of hypogonadism among men with weight loss related to human immunodeficiency virus infection who were receiving highly active antiretroviral therapy. Clin Infect Dis. 2000;31:1240–4.
19. Roubenoff R, Grinspoon S, Skolnik PR, et al. Role of cytokines and testosterone in regulating lean body mass and resting energy expenditure in HIV-infected men. Am J Physiol Endocrinol Metab. 2002;283:E138–45.
20. Kalyani RR, Gavini S, Dobs AS. Male hypogonadism in systemic disease. Endocrinol Metab Clin North Am. 2007;36:333–48.
21. Cubero JM, Domingo P, Sambeat M, et al. Prevalence of metabolic syndrome among human immunodeficiency virus-infected subjects is widely influenced by the diagnostic criteria. Metab Syndr Relat Disord. 2011;9:345–51.
22. Worm SW, Lundgren JD. The metabolic syndrome in HIV. Best Pract Res Clin Endocrinol Metab. 2011;25:479–86.
23. Hruz PW. Molecular mechanisms for insulin resistance in treated HIV-infection. Best Pract Res Clin Endocrinol Metab. 2011;25:459–68.
24. Selvin E, Feinleib M, Zhang L, et al. Androgens and diabetes in men: results from the Third National Health and Nutrition Examination Survey (NHANES III). Diabetes Care. 2007;30:234–8.
25. Rodriguez A, Muller DC, Metter EJ, et al. Aging, androgens, and the metabolic syndrome in a longitudinal study of aging. J Clin Endocrinol Metab. 2007;92:3568–72.
26. Monroe AK, Dobs AS, Xu X, et al. Sex hormones, insulin resistance, and diabetes mellitus among men with or at risk for HIV infection. J Acquir Immune Defic Syndr. 2011;58:173–80.
27. Calza L, Masetti G, Piergentili B, et al. Prevalence of diabetes mellitus, hyperinsulinaemia and metabolic syndrome among 755 adult patients with HIV-1 infection. Int J STD AIDS. 2011;22:43–5.
28. Apostolova N, Blas-Garcia A, Esplugues JV. Mitochondrial toxicity in HAART: an overview of in vitro evidence. Curr Pharm Des. 2011;17:2130–44.
29. Biglia A, Blanco JL, Martinez E, et al. Gynecomastia among HIV-infected patients is associated with hypogonadism: a case-control study. Clin Infect Dis. 2004;39:1514–9.
30. Rahim S, Ortiz O, Maslow M, et al. A case-control study of gynecomastia in HIV-1-infected patients receiving HAART. AIDS Read. 2004;14:23–4, 29–32, 35–40.
31. Mira JA, Lozano F, Santos J, et al. Gynaecomastia in HIV-infected men on highly active antiretroviral therapy: association with efavirenz and didanosine treatment. Antivir Ther. 2004;9:511–7.
32. Richardson D, Goldmeier D, Frize G. Letrozole versus testosterone. A single-center pilot study of HIV-infected men who have sex with men on highly active anti-retroviral therapy (HAART) with hypoactive sexual desire disorder and raised estradiol levels. J Sex Med. 2007;4:502–8.
33. Guaraldi G, Beggi M, Zona S, et al. Erectile dysfunction is not a mirror of endothelial dysfunction in HIV-infected patients. J Sex Med. 2012;9(4):111–1121.
34. Lallemand F, Salhi Y, Linard F, et al. Sexual dysfunction in 156 ambulatory HIV-infected men receiving highly active antiretroviral therapy combinations with and without protease inhibitors. J Acquir Immune Defic Syndr. 2002;30:187–90.
35. McLachlan RI, Allan CA. Defining the prevalence and incidence of androgen deficiency in aging men: where are the goal posts? J Clin Endocrinol Metab. 2004;89:5916–9.
36. Health Protection Agency. HIV in the United Kingdom: 2010. Nov 2010. http://www.hpa.org.uk/web/HPAweb&HPAwebStandard/HPAweb_C/1287145264558. Accessed 26 Apr 2012.

37. Kirk JB, Goetz MB. Human immunodeficiency virus in an aging population, a complication of success. J Am Geriatr Soc. 2009;57:2129–38.
38. Klein RS, Lo Y, Santoro N, et al. Androgen levels in older men who have or who are at risk of acquiring HIV infection. Clin Infect Dis. 2005;41:1794–803.
39. Desai S, Landay A. Early immune senescence in HIV disease. Curr HIV/AIDS Rep. 2010;7:4–10.
40. Cejtin HE, Kalinowski A, Bacchetti P, et al. Effects of human immunodeficiency virus on protracted amenorrhea and ovarian dysfunction. Obstet Gynecol. 2006;108:1423–31.
41. Schoenbaum EE, Hartel D, Lo Y, et al. HIV infection, drug use, and onset of natural menopause. Clin Infect Dis. 2005;41:1517–24.
42. Clark RA, Mulligan K, Stamenovic E, et al. Frequency of anovulation and early menopause among women enrolled in selected adult AIDS clinical trials group studies. J Infect Dis. 2001;184:1325–7.
43. Fantry LE, Zhan M, Taylor GH, et al. Age of menopause and menopausal symptoms in HIV-infected women. AIDS Patient Care STDS. 2005;19:703–11.
44. Cohan GR. HIV-associated hypogonadism. AIDS Read. 2006;16:341–5, 348, 352–4.
45. Cotter AG, Powderly WG. Endocrine complications of human immunodeficiency virus infection: hypogonadism, bone disease and tenofovir-related toxicity. Best Pract Res Clin Endocrinol Metab. 2011;25:501–15.
46. Ebeling PR. Androgens and osteoporosis. Curr Opin Endocrinol Diabetes Obes. 2010;17: 284–92.
47. Cooper OB, Brown TT, Dobs AS. Opiate drug use: a potential contributor to the endocrine and metabolic complications in human immunodeficiency virus disease. Clin Infect Dis. 2003;37 Suppl 2:S132–6.
48. Wisniewski AB, Brown TT, John M, et al. Hypothalamic-pituitary-gonadal function in men and women using heroin and cocaine, stratified by HIV status. Gend Med. 2007;4:35–44.
49. Centers for Disease Control (CDC). Revision of the CDC surveillance case definition for acquired immunodeficiency syndrome. MMWR Morb Mortal Wkly Rep. 1987;36 Suppl 1:3S–15.
50. Corcoran C, Grinspoon S. Treatments for wasting in patients with the acquired immunodeficiency syndrome. N Engl J Med. 1999;340:1740–50.
51. Dudgeon WD, Phillips KD, Carson JA, et al. Counteracting muscle wasting in HIV-infected individuals. HIV Med. 2006;7:299–310.
52. Kotler DP, Wang J, Pierson RN. Body composition studies in patients with the acquired immunodeficiency syndrome. Am J Clin Nutr. 1985;42:1255–65.
53. Ott M, Lembcke B, Fischer H, et al. Early changes of body composition in human immunodeficiency virus-infected patients: tetrapolar body impedance analysis indicates significant malnutrition. Am J Clin Nutr. 1993;57:15–9.
54. Grinspoon S, Corcoran C, Lee K, et al. Loss of lean body and muscle mass correlates with androgen levels in hypogonadal men with acquired immunodeficiency syndrome and wasting. J Clin Endocrinol Metab. 1996;81:4051–8.
55. Von Roenn JH, Armstrong D, Kotler DP, et al. Megestrol acetate in patients with AIDS-related cachexia. Ann Intern Med. 1994;121:393–9.
56. Oster MH, Enders SR, Samuels SJ, et al. Megestrol acetate in patients with AIDS and cachexia. Ann Intern Med. 1994;121:400–8.
57. Mwamburi DM, Gerrior J, Wilson IB, et al. Comparing megestrol acetate therapy with oxandrolone therapy for HIV-related weight loss: similar results in 2 months. Clin Infect Dis. 2004;38:895–902.
58. Beal JE, Olson R, Laubenstein L, Morales JO, Bellman P, Yangco B, et al. Dronabinol as a treatment for anorexia associated with weight loss in patients with AIDS. J Pain Symptom Manage. 1995;1:89–97.
59. Timpone JG, Wright DJ, Li N, et al. The safety and pharmacokinetics of single-agent and combination therapy with megestrol acetate and dronabinol for the treatment of HIV wasting

syndrome. The DATRI 004 Study Group. Division of AIDS Treatment Research Initiative. AIDS Res Hum Retroviruses. 1997;13:305–15.
60. Dejesus E, Rodwick BM, Bowers D, Cohen CJ, Pearce D. Use of dronabinol improves appetite and reverses weight loss in HIV/AIDS-infected patients. J Int Assoc Physicians AIDS Care (Chic). 2007;6:95–100.
61. Bhasin S, Javanbakht M. Can androgen therapy replete lean body mass and improve muscle function in wasting associated with human immunodeficiency virus infection? JPEN J Parenter Enteral Nutr. 1999;23:S195–201.
62. Schambelan M, Mulligan K, Grunfeld C, et al. Recombinant human growth hormone in patients with HIV-associated wasting. A randomized, placebo-controlled trial. Serostim Study Group. Ann Intern Med. 1996;125:873–82.
63. Lo JC, Mulligan K, Noor MA, et al. The effects of recombinant human growth hormone on body composition and glucose metabolism in HIV-infected patients with fat accumulation. J Clin Endocrinol Metab. 2001;86:3480–7.
64. Tai VW, Schambelan M, Algren H, et al. Effects of recombinant human growth hormone on fat distribution in patients with human immunodeficiency virus-associated wasting. Clin Infect Dis. 2002;35:1258–62.
65. Engelson ES, Glesby MJ, Mendez D, et al. Effect of recombinant human growth hormone in the treatment of visceral fat accumulation in HIV infection. J Acquir Immune Defic Syndr. 2002;30:379–91.
66. Storer TW, Woodhouse LJ, Sattler F, et al. A randomized, placebo-controlled trial of nandrolone decanoate in human immunodeficiency virus-infected men with mild to moderate weight loss with recombinant human growth hormone as active reference treatment. J Clin Endocrinol Metab. 2005;90:4474–82.
67. Moyle GJ, Daar ES, Gertner JM, et al. Growth hormone improves lean body mass, physical performance, and quality of life in subjects with HIV-associated weight loss or wasting on highly active antiretroviral therapy. J Acquir Immune Defic Syndr. 2004;35:367–75.
68. Gelato M, McNurlan M, Freedland E. Role of recombinant human growth hormone in HIV-associated wasting and cachexia: pathophysiology and rationale for treatment. Clin Ther. 2007;29:2269–88.
69. Waters D, Danska J, Hardy K, et al. Recombinant human growth hormone, insulin-like growth factor 1, and combination therapy in AIDS-associated wasting. A randomized, double-blind, placebo-controlled trial. Ann Intern Med. 1996;125:865–72.
70. Mulligan K, Tai VW, Schambelan M. Use of growth hormone and other anabolic agents in AIDS wasting. JPEN J Parenter Enteral Nutr. 1999;23:S202–9.
71. Ellis KJ, Lee PD, Pivarnik JM, et al. Changes in body composition of human immunodeficiency virus-infected males receiving insulin-like growth factor I and growth hormone. J Clin Endocrinol Metab. 1996;81:3033–8.
72. Lee PD, Pivarnik JM, Bukar JG, et al. A randomized, placebo-controlled trial of combined insulin-like growth factor I and low dose growth hormone therapy for wasting associated with human immunodeficiency virus infection. J Clin Endocrinol Metab. 1996;81:2968–75.
73. Rabkin JG, Wagner GJ, Rabkin R. A double-blind, placebo-controlled trial of testosterone therapy for HIV-positive men with hypogonadal symptoms. Arch Gen Psychiatry. 2000;57:141–7; discussion 155–6.
74. Coodley GO, Coodley MK. A trial of testosterone therapy for HIV-associated weight loss. AIDS. 1997;11:1347–52.
75. Grinspoon S, Corcoran C, Askari H, et al. Effects of androgen administration in men with the AIDS wasting syndrome. A randomized, double-blind, placebo-controlled trial. Ann Intern Med. 1998;129:18–26.
76. Grinspoon S, Corcoran C, Anderson E, et al. Sustained anabolic effects of long-term androgen administration in men with AIDS wasting. Clin Infect Dis. 1999;28:634–6.
77. Rabkin JG, Wagner GJ, Rabkin R. Testosterone therapy for human immunodeficiency virus-positive men with and without hypogonadism. J Clin Psychopharmacol. 1999;19:19–27.

78. Knapp PE, Storer TW, Herbst KL, et al. Effects of a supraphysiological dose of testosterone on physical function, muscle performance, mood, and fatigue in men with HIV-associated weight loss. Am J Physiol Endocrinol Metab. 2008;294:E1135–43.
79. Bhasin S, Storer TW, Javanbakht M, et al. Testosterone replacement and resistance exercise in HIV-infected men with weight loss and low testosterone levels. JAMA. 2000;283: 763–70.
80. Grinspoon S, Corcoran C, Parlman K, et al. Effects of testosterone and progressive resistance training in eugonadal men with AIDS wasting. A randomized, controlled trial. Ann Intern Med. 2000;133:348–55.
81. Kong A, Edmonds P. Testosterone therapy in HIV wasting syndrome: systematic review and meta-analysis. Lancet Infect Dis. 2002;2:692–9.
82. Edelstein D, Basaria S. Testosterone undecanoate in the treatment of male hypogonadism. Expert Opin Pharmacother. 2010;11:2095–106.
83. Dobs AS, Cofrancesco J, Nolten WE, et al. The use of a transscrotal testosterone delivery system in the treatment of patients with weight loss related to human immunodeficiency virus infection. Am J Med. 1999;107:126–32.
84. Miller K, Corcoran C, Armstrong C, et al. Transdermal testosterone administration in women with acquired immunodeficiency syndrome wasting: a pilot study. J Clin Endocrinol Metab. 1998;83:2717–25.
85. Bhasin S, Storer TW, Asbel-Sethi N, et al. Effects of testosterone replacement with a non-genital, transdermal system, androderm, in human immunodeficiency virus-infected men with low testosterone levels. J Clin Endocrinol Metab. 1998;83:3155–62.
86. Bhasin S, Parker RA, Sattler F, et al. Effects of testosterone supplementation on whole body and regional fat mass and distribution in human immunodeficiency virus-infected men with abdominal obesity. J Clin Endocrinol Metab. 2007;92:1049–57.
87. Scott JD, Wolfe PR, Anderson P, et al. Prospective study of topical testosterone gel (AndroGel) versus intramuscular testosterone in testosterone-deficient HIV-infected men. HIV Clin Trials. 2007;8:412–20.
88. Berger JR, Pall L, Hall CD, et al. Oxandrolone in AIDS-wasting myopathy. AIDS. 1996;10:1657–62.
89. Fisher A, Abbatiola M. The effects of oxandrolone on body weight and composition in patients with HIV-associated weight loss. Int Conf AIDS. 1998;12:844.
90. Grunfeld C, Kotler DP, Dobs A, et al. Oxandrolone in the treatment of HIV-associated weight loss in men: a randomized, double-blind, placebo-controlled study. J Acquir Immune Defic Syndr. 2006;41:304–14.
91. Gold J, High HA, Li Y, et al. Safety and efficacy of nandrolone decanoate for treatment of wasting in patients with HIV infection. AIDS. 1996;10:745–52.
92. Sattler FR, Jaque SV, Schroeder ET, et al. Effects of pharmacological doses of nandrolone decanoate and progressive resistance training in immunodeficient patients infected with human immunodeficiency virus. J Clin Endocrinol Metab. 1999;84:1268–76.
93. Hengge UR, Stocks K, Wiehler H, et al. Double-blind, randomized, placebo-controlled phase III trial of oxymetholone for the treatment of HIV wasting. AIDS. 2003;17:699–710.
94. Bravo P, Edwards A, Rollnick S, et al. Tough decisions faced by people living with HIV: a literature review of psychosocial problems. AIDS Rev. 2010;12:76–88.
95. Grinspoon S, Corcoran C, Stanley T, et al. Effects of hypogonadism and testosterone administration on depression indices in HIV-infected men. J Clin Endocrinol Metab. 2000;85: 60–5.
96. Rabkin JG, Wagner GJ, McElhiney MC, et al. Testosterone versus fluoxetine for depression and fatigue in HIV/AIDS: a placebo-controlled trial. J Clin Psychopharmacol. 2004;24: 379–85.
97. Gutierrez F, Masia M. The role of HIV and antiretroviral therapy in bone disease. AIDS Rev. 2011;13:109–18.

98. Bhasin S, Cunningham GR, Hayes FJ, et al. Testosterone therapy in men with androgen deficiency syndromes: an Endocrine Society clinical practice guideline. J Clin Endocrinol Metab. 2010;95:2536–59.
99. Clay PG, Lam AI. Testosterone replacement therapy for bone loss prevention in HIV-infected males. Ann Pharmacother. 2003;37:582–5.
100. Fairfield WP, Finkelstein JS, Klibanski A, et al. Osteopenia in eugonadal men with acquired immune deficiency syndrome wasting syndrome. J Clin Endocrinol Metab. 2001;86: 2020–6.

Chapter 10
Treatment Options for Testosterone Replacement Therapy

Mohit Khera

Over the past half decade, the testosterone replacement therapy (TRT) market has grown rapidly. Exogenous testosterone in various forms is now the second fastest growing medication prescribed in the USA. From 2005 to 2009, spending on testosterone jumped by 115.5%, and the number of prescriptions filled increased by 64.5%. While only two testosterone gel therapies had been available for the previous 10 years, 2011 saw three new gels enter the TRT market (Fig. 10.1).

There are many reasons for the rapid growth of the TRT market. We have an aging population, with the number of men 65 and older in the USA increasing two to three times faster than the number of men younger than 65. Furthermore, recent data demonstrate an increased association between poor general health, possible mortality, and low serum testosterone levels. There is now less concern for the development of prostate cancer after TRT, making it a more attractive treatment option. Finally, new drugs are entering the TRT market with increased promotion, marketing, and direct-to-consumer advertising and are driving market growth.

Brief History of Testosterone Supplementation

The beneficial effects of "testosterone" have been suspected for thousands of years. In 2000 BC, the ancient Indian manuscripts described the ingestion of testicular tissue for the treatment of male impotence. The ancient Egyptians also described the medicinal powers of the testis. In 1889, Charles Brown-Sequard injected himself with an extract of crushed canine and guinea pig testes, and reported improvements in his urinary stream, intellect, and erectile function. In 1920, Serge Vornoff

M. Khera, MD, MBA, MPH (✉)
Scott Department of Urology, Baylor College of Medicine,
6620 Main #1325, Houston, TX 77030, USA
e-mail: mkhera@bcm.edu

W.J.G. Hellstrom (ed.), *Androgen Deficiency and Testosterone Replacement: Current Controversies and Strategies*, Current Clinical Urology,
DOI 10.1007/978-1-62703-179-0_10,

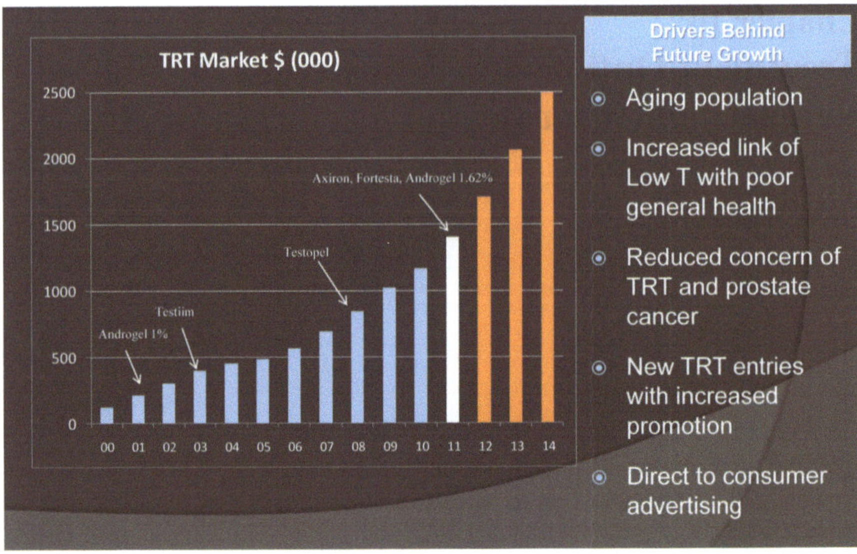

Fig. 10.1 TRT market overview

completed the first testicular tissue transplant from chimpanzee to human [1]. In 1941, Butenandt and Ruzicka first described the synthesis of testosterone. In the 1940s, the first subdermal testosterone implants were introduced, and 10 years later the first testosterone esters followed. These esters are the basis of the intramuscular injections that we still use today. In the 1970s, testosterone undeconoate (TU; Andriol Testocaps®, Organon) became available outside the USA. Currently, TU is in Phase III trials in the USA. In 1994, the first transdermal testosterone was introduced as a patch, known as Testoderm (Alza Corporation), and in 2008, the FDA first approved subdermal testosterone implants.

Rationale for Testosterone Replacement

Androgen deficiency is a serious problem in the older male population, thought to affect approximately 1 in 200 men [2]. Mulligan et al. demonstrated that roughly 39% of men over the age of 45 had an androgen deficiency, using a threshold testosterone level <300 ng/dL [3]. However, not all of these men were symptomatic. In fact, Araujo et al. found that the prevalence of *symptomatic* androgen deficiency was approximately 4% in men under the age of 50 and 8% in men over the age of 50 [4]. The European Male Aging Study noted that only 2% of men between the ages of 40 and 79 had symptomatic androgen deficiency. Thus, clinicians must exercise caution in selecting patients who should receive treatment.

We recognize that beginning at approximately 20–30 years of age, males experience a decline in total testosterone and free testosterone levels of 0.4% and 1.3% per year, respectively [5]. The common symptoms observed with this decline in serum androgens have not been fully defined, as have the symptoms of declining estrogen levels in women. Signs and symptoms of hypogonadism include changes in bone mineral density [6, 7], muscle strength [8, 9], cognition [10], body composition [11], and sexual function [12].

Testosterone Production and Secretion

In order to understand how different testosterone therapies function, one must understand the hypothalamic–pituitary–gonadal axis. The production of testosterone is regulated by luteinizing hormone (LH) from the anterior pituitary. In turn, LH secretion is regulated by gonadotropin-releasing hormone (GnRH), which is secreted from the hypothalamus. GnRH is secreted in a pulsatile fashion and is significantly elevated at the time a boy enters puberty. This elevation in GnRH, and subsequent increase in LH, results in a surge in testosterone and the subsequent development of secondary male sexual characteristics and spermatogenesis.

Testosterone is synthesized by the Leydig cells within the testicles. The testes are not able to store testosterone adequately on their own or convert testosterone into its more potent form, dihydrotestosterone (DHT). This is because the testes lack the enzyme, 5-alpha-reductase, which converts testosterone into DHT, its much more potent form. In order to compensate for this, the testes have high levels of androgen binding protein, which is produced by the Sertoli cells and helps to maintain high intratesticular levels of testosterone [13].

Testosterone levels vary on the basis of a diurnal rhythm, with the highest levels of circulating testosterone occurring in the early morning hours. Studies have demonstrated that the diurnal rhythm of total testosterone that is observed in young men is markedly reduced or absent in older, yet otherwise healthy, men suggesting that these altered diurnal patterns may be a consequence of normal aging [14].

Approximately 2% of circulating testosterone within the blood is unbound and freely enters cells to exert a metabolic effect [15, 16]. The amount of bioavailable testosterone is the sum of the free testosterone and the albumin-bound testosterone [15]. The distinction between bioavailable testosterone and total testosterone (bioavailable plus SHBG bound) may be important when levels of testosterone are being measured in the hypogonadal male. SHBG is synthesized in the liver, and its levels can be increased by hyperthyroidism and cirrhosis, and decreased by hypothyroidism, acromegaly, and obesity [15, 16]. SHBG levels also increase as men age, and these increases can lead to variability in testosterone levels and their impact on biological functions [15, 16].

Raising Endogenous Testosterone

Use of exogenous testosterone should be discouraged in hypogonadal patients who desire to protect their future fertility potential. A more appropriate approach in such patients is to intrinsically increase their endogenous testosterone. There are several ways to accomplish this; however, all of the options, except for HCG injections, are considered off label.

One option is to initiate use of clomid, 50 mg every other day or 25 mg every day for 3–6 months. Clomiphene citrate is a selective estrogen receptor modulator. A trans isomer of clomiphene citrate is currently in clinical trials in the USA for the treatment of secondary male hypogonadism. Because clomiphene citrate is an estrogen receptor blocker and raises serum FSH and LH levels, it is less effective in raising serum testosterone levels when a patient's LH and FSH levels are already elevated, as in patients with primary testis failure.

Another method of increasing endogenous testosterone is the use of Arimidex (AstraZeneca). Arimidex is an aromatase inhibitor that blocks the conversion of testosterone to estradiol. This action is especially useful in obese patients who tend to have increased amounts of aromatization occurring in their fatty tissues. In adults, there is concern that long-standing suppression of estradiol may increase the risk of osteoporosis and osteopenia, and lead to joint pain. However, this has not been the case in adolescent young men.

BHCG injections are an alternate method, used to increase endogenous serum testosterone levels. HCG is an LH analog that stimulates Leydig cell production of testosterone. Intramuscular injections vary from two to three times per week and from 1,500 to 3,000 units and sometimes higher. While HCG injections may be beneficial in raising serum testosterone levels and preserving fertility, HCG injections can be expensive and are not covered by insurance. The invasive nature of this medication can deter many patients.

Methods of Exogenous Androgen Replacement

The primary goal of TRT is to restore normal physiologic concentrations of testosterone. The choice method for TRT depends on availability, safety, cost, tolerability, efficacy, and patient and physician preference. A summary of the currently available testosterone delivery formulations is listed in Table 10.1.

Intramuscular Injections

Injectable testosterone first became available in the USA in the 1950s. Intramuscular administration of exogenous testosterone offers a cost-effective and efficacious method of androgen replacement. Peak serum concentrations are achieved within 72 h, and injections are administered every 7–21 days, depending on symptom control, type of steroid, and androgen response in individual patients. Testosterone cypionate and testosterone enanthate must be injected every 2–4 weeks and testosterone propionate two to three times per week. Among the drawbacks of intramuscular

Table 10.1 TRT options and treatment dosages

Testosterone enanthate or cypionate IM	100 mg q week
	200 mg q2 weeks
	60 mg 2×/week
Testosterone patch	2–4 g (1–2 patches) qhs
Testosterone liquid	60–120 mg qd
Testosterone gel	5–10 g qd
Bioadhesive buccal testosterone	30 mg q12h
Testosterone pellets	75 g × 10 q4–6 m
Testosterone undecanoate	Not available USA

Based on data from Bhasin S, Cunningham GR, Hayes FJ, et al. "Testosterone Therapy in Adult Men with Androgen Deficiency Syndromes: An Endocrine Society Clinical Practice Guideline" J Clin Endocrinol Metab 2006;91:1995–2010

injections is the lack of circadian release of testosterone, causing transient supraphysiologic levels in the first 2–3 days after an injection, followed by a steady decline to subphysiologic levels just before the next injection. The sudden decline of testosterone towards the end of 2–3 week interval is also known as the "testosterone crash," and is associated with a sudden and severe recurrence of hypogonadal symptoms. Currently available preparations include testosterone enanthate, cypionate, propionate, and undecanoate (not available in the USA).

The injectable testosterones have esterification of testosterone at the 17B-hydroxy position which makes the compound more hydrophobic, and allows for action of a longer duration. The addition of oil as a carrier of the testosterone molecule also increases the half-life. Complications specific to injectable testosterones include hematomas, ecchymosis, and pain at the injection site [17, 18].

Due to peaks and troughs following testosterone injection, mood swings can occur [17]. Furthermore, patients receiving injectable testosterone are more susceptible to erythrocytosis. One study found a rise in hematocrit in 24% of patients after injections of testosterone cypionate without any adverse effects [19]. Older patients are much more likely to develop erythrocytosis, and caution and close follow-up should be taken in this population.

Long-Acting Testosterone Injections

Nebido® (Schering AG), a long-acting injectable form of TU, was approved in Europe in 2004. It still is pending FDA approval in the USA. Nebido is an intramuscular testosterone that is injected every 10–12 weeks (after a 6-week loading dosage following the first injection). The common dosage used in Europe is 1,000 mg of testosterone; the dose now being evaluated for the United States is 750 mg. The testosterone is administered intramuscularly in 4 cc of castor oil, which enhances its ability to release slowly [20].

Transdermal Testosterone

Currently, most TRT users in the USA are receiving some form of transdermal gel therapy. In 2006, roughly 70% of TRT prescriptions were for transdermal gels; 17% were using testosterone injections; 10% were using testosterone patches; and 3% were using some other form of testosterone supplementation, such as oral formulations. Options for transdermal androgen replacement include adhesive skin patches or gel applications.

Testosterone Patches

Patches were first introduced in 1994 as scrotal patches (Testoderm) but were discontinued because of scrotal irritation. Scrotal patches were also associated with increased levels of DHT because of the scrotum's high levels of 5-alpha-reductase. Soon after non-scrotal testosterone patches (Androderm and Andropatch) were introduced as 2.5 and 5 mg patch doses. Patches could be combined for a maximum dose of 10 mg. Patches are typically applied at night to the back, abdomen, thighs, or upper arms. The 5-mg patch delivers approximately 5 mg/day of testosterone, which is the average amount of testosterone produced by a man in a day. While gels can achieve a physiologic steady state with daily application, patches recreate the physiologic circadian release of androgens.

Drawbacks of skin patches include their visibility and more frequent skin reactions. In fact, up to 37% of men can develop some type of skin irritation after applying testosterone patches. Some reports suggest pretreating under the patch, with 1% triamcinolone acetonide, can decrease the risk of dermatitis without compromising testosterone absorption [21]. Other adverse reactions include skin induration, vesicle formation, allergic contact dermatitis, headaches, and depression.

Testosterone Gels/Liquids

Testosterone gels were first introduced into the American market in 2000. Daily testosterone gel preparations (applied daily) represent an effective method of hormonal replacement with a respectable safety profile, and they have quickly become the first-line therapy for the majority of men receiving testosterone replacement. Currently, there are six different gels available in the USA. These include AndroGel (1%), AndroGel (1.62%), Testim, Axiron, Foresta, and Andractim. Only Axiron is considered a liquid. Each of these gels and liquids has different characteristics, and they are applied in various body locations.

Androgel is available in two different concentrations: the original 1% and the newly introduced 1.62%. There are some differences between them (Table 10.2).

Table 10.2 Comparison of androgen 1% and androgen 1.62%

	Androgel 1%	Androgel 1.62%
Application site	Shoulders, upper arms, abdomen	Shoulders, upper arms
Starting pumps	4	2
Starting dose	50 mg	40.5 mg
Maximum dose	100 mg	81 mg
Time to swim or shower	6 h	2 h

For example, the starting dose of Androgel 1% is 50 mg of testosterone and for Androgel 1.62% is 40.5 mg of testosterone. The maximum doses of testosterone in Androgel 1 and 1.62% are 100 mg and 81 mg, respectively. A patient can swim or shower 6 h after applying Androgel 1% and 2 h after using Androgel 1.62%. Finally, unlike Androgel 1%, Androgel 1.62% is contraindicated for application to the abdomen. Adverse effects specific for Androgel include acne (1–8%), headaches (4%), emotional lability (3%), nervousness, and gynecomastia.

Testim is a 1% gel in which each tube contains 50 mg of testosterone. The starting dose is 1 tube, but some patients are titrated to two tubes, or 100 mg of testosterone daily. Testim is applied to the shoulders and upper arms. In several studies, Testim has been shown to have very good skin penetration and serum testosterone levels [22, 23]. This result is due primarily to the emollient, pentadecalactone, which is often found in many aftershaves and colognes. Patients can swim or shower 2 h after applying Testim.

Fortesta is a 2% gel, introduced into the TRT market after being used throughout Europe. This gel is applied to the inner thighs starting with a dose of 40 mg. Each pump is 10 mg, and a total of two pumps are placed on each inner thigh. The dose can be titrated to 70 mg, and patients can swim or shower after 2 h. The major side effect of Fortesta is skin reaction in 16% of patients.

Axiron 2% has also been introduced recently into the TRT market. It is considered a liquid, and not a gel. This liquid is placed under the axilla, at a starting dose of 60 mg which can be titrated to 120 mg. Each pump contains 30 mg of testosterone. The patient applies his deodorant before the liquid. Adverse effects of Axiron include application site erythema and irritation in 7–8%, headaches (6%), erythrocytosis (7%), diarrhea (4%), and vomiting (4%).

The major benefits of the gels are that they can be stopped at any time and doses can be easily titrated to meet a patient's needs. Patients who do not respond to one form of gel therapy may try another gel or liquid, as this may improve serum testosterone levels. Patients receiving gel or liquid therapy may experience higher levels of DHT, because the dermis contains 5-alpha-reductase, which is responsible for converting testosterone into DHT. Currently, all testosterone liquids and gels in the USA contain an FDA black box warning for the risk of transference to women or children. Thus, the gel or liquid needs to be washed off before there is any skin-to-skin contact between the testosterone application site and another person.

Buccal Formulations

Striant is currently the only buccal formulation of testosterone available in the USA. It is applied twice daily to the buccal mucosa above the upper incisor teeth. Each tablet contains 30 mg of testosterone, and tablets must be reapplied every 12 h. The application site needs to be rotated to minimize local gum irritation, which occurs in >10% of patients. Striant avoids first-pass hepatic metabolism, by its direct absorption through the buccal mucosa and then into the circulation. This delivery system allows for a rapid absorption of testosterone, which reaches a peak level in 30 min [24]. The main side effects from Striant are gum irritation including edema, gingivitis, blistering, and inflammation. Altered taste has also been described by some patients.

Oral Testosterone Preparations

Orally administered testosterone is almost completely inactivated by first-pass metabolism by the liver. This can result in significant hepatotoxicity and lipid profile abnormalities. Newer alkylated testosterone agents, such as with oral TU, have been developed to avoid hepatotoxicity. TU must be taken with a fatty meal because it is absorbed into the lymphatic system and travels through the circulation and avoids the liver. The average dose is 120–240 mg/day and it peaks in the circulation at 2–6 h after dosing [25].

Subcutaneous Testosterone Pellets

Subcutaneous pellets have been available for decades but were approved only recently (2008) by the FDA Testopel testosterone pellets (Testopel®, Slate Pharmaceuticals) are currently the only long-acting testosterone supplementation available in the USA. The pellets are placed in the fat layer under the skin of the buttock or abdomen in a simple office procedure. The pellets typically dissolve over 3–6 months and need to be replaced. The potential benefits for Testopel is there is no risk for transference, improved patient compliance, and convenience for not having to apply the agent daily. Potential risks of Testopel include bleeding, bruising, infection, expulsion of the pellets, and pain. A study by Cavender et al. found that 1.3% of Testopel insertions precipitated an adverse event, such as pruritus, erythema, infection, or foreign body reaction [26]. The infection rate in this series was 0.3%. In a recent multi-institutional study, Testopel provided sustained levels of testosterone for at least 3 months and up to 6 months in men with some form of testosterone deficiency [27]. Implantation of more than eight pellets achieves optimal results with respect to peak mean testosterone level and duration of effect. Testosterone pellets are generally well tolerated.

Adverse Side Effects of Testosterone Replacement Therapy

Absolute contraindications to TRT include untreated or residual prostate or breast cancer. Serious adverse side effects secondary to exogenous testosterone administration are relatively uncommon in a medically supervised program. Adverse effects appear to be particularly important in elderly patients and they are often dependent on the method of testosterone supplementation. Reported adverse effects include:

- Polycythemia
- Prostate enlargement or exacerbation of benign prostatic hypertrophy (BPH)
- Gynecomastia
- Hepatotoxicity
- Lipid profile abnormalities
- Impaired sperm production and infertility
- Fluid retention
- Sleep apnea
- Acne or oily skin
- Hair loss from the scalp

Often these adverse effects can be minimized or negated altogether by dose adjustment, switching to an alternative form of therapy, or discontinuation of androgen supplementation.

Summary

TRT is growing exponentially in the USA and newer treatment modalities are rapidly entering the market. While most patients are currently using gel therapy, other treatments such as subcutaneous testosterone pellets and testosterone undeconoate are beginning to gain interest among patients. Exogenous testosterone should not be used in men who desire to initiate a pregnancy and methods of raising endogenous testosterone levels should be considered.

References

1. Miller NL, Fulmer BR. Injection, ligation and transplantation: the search for the glandular fountain of youth. J Urol. 2007;177:2000–5.
2. Morales A, Tenover JL. Androgen deficiency in the aging male: when, who, and how to investigate and treat. Urol Clin North Am. 2002;29:975–82.
3. Mulligan T, Frick MF, Zuraw QC, Stemhagen A, McWhirter C. Prevalence of hypogonadism in males aged at least 45 years: the HIM study. Int J Clin Pract. 2006;60:762–9.
4. Araujo AB, Esche GR, Kupelian V, et al. Prevalence of symptomatic androgen deficiency in men. J Clin Endocrinol Metab. 2007;92:4241–7.

5. Wu FC, Tajar A, Pye SR, et al. Hypothalamic-pituitary-testicular axis disruptions in older men are differentially linked to age and modifiable risk factors: the European Male Aging Study. J Clin Endocrinol Metab. 2008;93:2737–45.
6. Kiratli BJ, Srinivas S, Perkash I, Terris MK. Progressive decrease in bone density over 10 years of androgen deprivation therapy in patients with prostate cancer. Urology. 2001;57:127–32.
7. Stoch SA, Parker RA, Chen L, et al. Bone loss in men with prostate cancer treated with gonadotropin-releasing hormone agonists. J Clin Endocrinol Metab. 2001;86:2787–91.
8. Bhasin S, Storer TW, Berman N, et al. Testosterone replacement increases fat-free mass and muscle size in hypogonadal men. J Clin Endocrinol Metab. 1997;82:407–13.
9. Murray MP, Gardner GM, Mollinger LA, Sepic SB. Strength of isometric and isokinetic contractions: knee muscles of men aged 20 to 86. Phys Ther. 1980;60:412–9.
10. Moffat SD, Zonderman AB, Metter EJ, Blackman MR, Harman SM, Resnick SM. Longitudinal assessment of serum free testosterone concentration predicts memory performance and cognitive status in elderly men. J Clin Endocrinol Metab. 2002;87:5001–7.
11. Katznelson L, Finkelstein JS, Schoenfeld DA, Rosenthal DI, Anderson EJ, Klibanski A. Increase in bone density and lean body mass during testosterone administration in men with acquired hypogonadism. J Clin Endocrinol Metab. 1996;81:4358–65.
12. Davidson JM, Chen JJ, Crapo L, Gray GD, Greenleaf WJ, Catania JA. Hormonal changes and sexual function in aging men. J Clin Endocrinol Metab. 1983;57:71–7.
13. van Roijen JH, Ooms MP, Weber RF, Brinkmann AO, Grootegoed JA, Vreeburg JT. Comparison of the response of rat testis and accessory sex organs to treatment with testosterone and the synthetic androgen methyltrienolone (R1881). J Androl. 1997;18:51–61.
14. Bremner WJ, Vitiello MV, Prinz PN. Loss of circadian rhythmicity in blood testosterone levels with aging in normal men. J Clin Endocrinol Metab. 1983;56:1278–81.
15. Braunstein GD. Testes. In: Greenspan FS, Strewler GJ, editors. Basic & Clinical Endocrinology. 6th ed, Chapter 12. Stamford, CT: Appleton & Lange; 1997. p. 422–52.
16. Petak SM, Nankin HR, Spark RF, Swerdloff RS, Rodriguez-Rigau LJ. American Association of Clinical Endocrinologists Medical Guidelines for clinical practice for the evaluation and treatment of hypogonadism in adult male patients—2002 update. Endocr Pract. 2002;8:440–56.
17. Dobs AS, Meikle AW, Arver S, Sanders SW, Caramelli KE, Mazer NA. Pharmacokinetics, efficacy, and safety of a permeation-enhanced testosterone transdermal system in comparison with bi-weekly injections of testosterone enanthate for the treatment of hypogonadal men. J Clin Endocrinol Metab. 1999;84:3469–78.
18. Edelstein D, Dobs A, Basaria S. Emerging drugs for hypogonadism. Expert Opin Emerg Drugs. 2006;11:685–707.
19. Sih R, Morley JE, Kaiser FE, Perry 3rd HM, Patrick P, Ross C. Testosterone replacement in older hypogonadal men: a 12-month randomized controlled trial. J Clin Endocrinol Metab. 1997;82:1661–7.
20. Behre HM, Abshagen K, Oettel M, Hubler D, Nieschlag E. Intramuscular injection of testosterone undecanoate for the treatment of male hypogonadism: phase I studies. Eur J Endocrinol. 1999;140:414–9.
21. Cummings DE, Kumar N, Bardin CW, Sundaram K, Bremner WJ. Prostate-sparing effects in primates of the potent androgen 7alpha-methyl-19-nortestosterone: a potential alternative to testosterone for androgen replacement and male contraception. J Clin Endocrinol Metab. 1998;83:4212–9.
22. Grober ED, Khera M, Soni SD, Espinoza MG, Lipshultz LI. Efficacy of changing testosterone gel preparations (Androgel or Testim) among suboptimally responsive hypogonadal men. Int J Impot Res. 2008;20:213–7.
23. Marbury T, Hamill E, Bachand R, Sebree T, Smith T. Evaluation of the pharmacokinetic profiles of the new testosterone topical gel formulation, Testim, compared to AndroGel. Biopharm Drug Dispos. 2003;24:115–20.
24. Kim S, Snipes W, Hodgen GD, Anderson F. Pharmacokinetics of a single dose of buccal testosterone. Contraception. 1995;52:313–6.

25. Gooren LJ. A ten-year safety study of the oral androgen testosterone undecanoate. J Androl. 1994;15:212–5.
26. Cavender RK, Fairall M. Subcutaneous testosterone pellet implant (Testopel) therapy for men with testosterone deficiency syndrome: a single-site retrospective safety analysis. J Sex Med. 2009;6:3177–92.
27. McCullough AR, Khera M, Goldstein I, Hellstrom WJ, Morgentaler A, Levine LA. A multi-institutional observational study of testosterone levels after testosterone pellet (testopel(R)) insertion. J Sex Med. 2012;9:594–601.

Chapter 11
Alternate Therapies for Testosterone Replacement

Ege Can Serefoglu, Ahmet Gokce, Wayne J.G. Hellstrom, and André T. Guay

Introduction

Male hypogonadism is an increasingly recognized clinical entity among the elderly population [1]. It can present with a variety of clinical manifestations and significantly affects health-related quality of life [2]. Since hypogonadism may result in multiple effects, including sexual dysfunction [3–6], decreased bone mineral density (BMD) [7], anemia [8], diminished strength due to muscle atrophy [9], increased cardiovascular risk [10], worsening cognitive function [11, 12], and development of depression [13], recent guidelines recommend hormone replacement when indicated [14].

Most hypogonadal men are treated with exogenous testosterone (T), which restores androgens to normal levels [15]. Oral, intramuscular, gel or subcutaneous forms of T replacement are currently the preferred treatment method for male hypogonadism [16, 17]. However, T replacement can have side effects [18, 19]. Exogenous T treatment suppresses follicle stimulating hormone (FSH) and luteinizing hormone (LH) release from the pituitary gland through a negative feedback mechanism. This may fail to restore sexual function, induce testicular atrophy and cause infertility [20, 21]. Moreover, excess T can be converted to estradiol and

E.C. Serefoglu, MD (✉)
Department of Urology, Tulane University Medical Center, 1430 Tulane Avenue SL-42, New Orleans, LA, USA
e-mail: eserefog@tulane.edu

A. Gokce, MD
Department of Urology, Tulane University School of Medicine, New Orleans, LA, USA

W.J.G. Hellstrom, MD, FACS
Department of Urology, Tulane University Medical Center, New Orleans, LA, USA

A.T. Guay, MD, FACP, FACE
Center for Sexual Function, Harvard Medical School, Peabody, MA, USA

W.J.G. Hellstrom (ed.), *Androgen Deficiency and Testosterone Replacement: Current Controversies and Strategies*, Current Clinical Urology,
DOI 10.1007/978-1-62703-179-0_11,

cause gynecomastia [22]. T replacement therapy is contraindicated in patients with prostate or breast cancer and may cause elevated PSA or hematocrit watson lower urining symptoms secondary to benign hyperplasia (BPH) aggravate untreated congestive heart failure or sleep apnea [14], especially if supraphysiological T levels are achieved. Residual T on the skin may be transferred to a spouse or child by contact contamination when transdermal gels are used, even 8 h after application, which may cause virilization [23, 24].

Since T replacement therapies may not be suitable for every case of hypogonadism, in part due to the aforementioned reasons, other treatment options may be offered [25]. For instance, if a young man with hypogonadotropic hypogonadism wishes to preserve his fertility, clomiphene citrate (CC), human chorionic gonadotropin (hCG), human menopausal gonadotropin, or recombinant FSH therapies can be prescribed [16]. In a recent study conducted by the American Urological Association, approximately 400 urologists were questioned on their empirical medical therapy of choice for idiopathic male infertility [26]. Although most of the participants noted that they prefered CC, hCG, and anastrozole treatments, 25% of these urologists stated that they would treat such an infertile man with T replacement, despite its contraceptive effect [21, 27].

This study clearly demonstrates a need for a better understanding of alternative therapies for T replacement. These various therapies are reviewed in this chapter.

Clomiphene Citrate

CC is a selective estrogen (E) receptor modulator and has traditionally been used for the treatment of ovulatory disorders in women. It is a weak E receptor antagonist, which blocks the negative feedback mechanism of circulating estradiol on the hypothalamic level, which in turn releases more GnRH [28]. Increased GnRH levels increase FSH and LH release, which stimulates T production and spermatogenesis by the testes [29].

In spite of the promising role of CC in treating hypogonadism, there are limited data regarding its efficacy. In one study, Katz et al. demonstrated the long-term (19 months) safety and efficacy of CC therapy in 89 young hypogonadal men (age 22–37), concluding that this treatment can be an alternative to T replacement therapy in men wishing to preserve their fertility [30]. Tenover et al. [29] compared the increase in T levels with 100 mg daily CC treatment in five healthy young men (<34 years old) and five healthy older men (>65 years old) and demonstrated a more robust response in the younger group.

In another study, Guay et al. [20] evaluated the hormone levels and sexual function in 17 men with erectile dysfunction (ED) who had secondary hypogonadism. This small double-blind, placebo-controlled, crossover study consisted of treatment with CC and a placebo for 2 months each. The authors demonstrated that LH, FSH, total and free T levels significantly increased; however, sexual function, as monitored by questionnaires and nocturnal penile tumescence and rigidity testing, did not improve except for some limited parameters in younger and healthier men. These results confirmed that there can be a functional secondary hypogonadism, but correction of the hormonal status does not necessarily reverse the associated ED back to normal. This suggests that a closer scrutiny of claims of cause and effect relationships between hypogonadism and ED is needed [20].

The same author evaluated 178 men with secondary hypogonadism and ED who were treated with CC for 4 months. Sexual function improved in 75%, but was corrective in only 39%, in spite of an increase in both LH and free T levels in all patients. The authors explained that treatment responses decreased significantly with older age and the presence of concomitant medical conditions such as diabetes, hypertension, coronary artery disease, and multiple medication use [31]. Daily low dose (25 mg) CC treatment was effective in 36 hypogonadal men, as the T/E ratio increases and improves the clinical symptoms of hypogonadism [22, 32]. Moreover, CC treatment is proven to be more effective in increasing T levels, despite its lower cost compared to T gels ($83 vs. $265–270/month) [33].

There are conflicting reports regarding the efficacy of CC in patients with pituitary disorders. CC was not effective in nine patients who were followed for nonfunctional pituitary adenomas. Only one of those patients had an increase in T levels and improvement in ED status [34]. This might be due to the fact that pituitary adenomas destroy the pituitary tissue, making these tissues unresponsive to CC stimulation. On the other hand, 12 weeks of CC treatment was found to be effective in 14 hyperprolactinemia patients. Of those patients, ten had increased T levels along with an increase in FSH, LH, and E. Moreover, statistically significant increase in sperm motility was detected in six patients who had asthenospermia prior to starting therapy [35].

Due to its minimal side effects, oral application, and lack of adverse effect on spermatogenesis, CC has a promising role in the treatment of hypogonadism, especially among young men who want to preserve their fertility. The low cost of this treatment and its ability to not cause testicular atrophy can be another advantage of this over alternative T replacement therapies.

Enclomiphene Citrate

Enclomiphene Citrate (Androxal™) is a trans-stereoisomer of CC and increases T production in the testes [36]. Because of the antagonistic effects of enclomiphene, the drug has the potential to increase serum T levels in men with secondary hypogonadism by restoring physiological endogenous testosterone secretion, while maintaining testicular volume and, potentially, spermatogenesis. In the FDA Phase II clinical trials conducted to date, enclomiphene demonstrated significant efficacy in the physiological restoration of testosterone levels in males with secondary hypogonadism. The compound also exhibited an unanticipated favorable effect on fasting plasma glucose. This result has been accompanied by rapidly accumulating evidence from other researchers for a bidirectional relationship between low serum testosterone and obesity/metabolic syndrome in men. Short-term clinical safety data for enclomiphene have been satisfactory and equivalent to safety data for T gels and placebo. Enclomiphene demonstrates benefits in the management of secondary hypogonadism associated with obesity, metabolic syndrome and possibly infertility. This agent will need to undergo Phase III, placebo-controlled, randomized clinical trials to document any effects in these noted conditions [37].

Anastrozole

Anastrozole is a nonsteroidal aromatase inhibitor. Its half life is longer than 24 h which enables daily dosage [38]. In a group of men with idiopathic hypogonadotropic hypogonadism and premature ejaculation, a 2 week anastrozole treatment restored the T, LH, and E levels, without any change on premature ejaculation [39]. It blocks the aromatization of T to estradiol. Since there are as many estradiol receptors in the hypothalamus and pituitary as there are T receptors, the body senses a deficiency and stimulates LH and FSH to correct the situation, raising the T level.

In a randomized, double-blind, placebo-controlled trial, a 1-year anastrozole 1 mg daily treatment increased free and total T levels in 88 older hypogonadal men (>60 years). Estradiol levels were decreased; however, no changes were observed either in body composition or strength [40]. In another study, anastrozole was combined with T in epileptic hypogonadal men. This combination therapy was found to be more effective in restoring the sexual function compared to T combined with placebo (72.2% vs. 47.4%) [41]. The frequency of epileptical seizures also significantly reduced in both groups.

The efficacy of anastrozole treatment seems to decrease with advancing age. In a double-blind study, 37 elderly men (aged 62–74 years) with serum T levels lower than 350 ng/dL were randomized to receive anastrozole 1 mg daily, twice weekly or daily placebo treatment [42]. At the end of 12 weeks of treatment, T levels significantly increased along with a decrease in E levels. However, no change was observed in sexual function (assessed with International Index of Erectile Function) or quality of life (assessed with MOS Short-Form Health Survey) in those patients. This treatment did not affect the serum levels of fasting lipid profile or insulin sensitivity, suggesting that it does not have an adverse effect on the cardiovascular system [43].

These authors also assessed the impact of anastrozole on bone metabolism by measuring serum and urine biochemical markers of bone turnover, serum osteoprotegerin, and total body BMD at baseline and week number 12 [44]. These researchers did not observe any increases in biochemical markers for bone resorption, bone formation, serum osteoprotegerin, or total body BMD. They concluded that anastrozole did not have an adverse effect on bone metabolism in a short treatment period.

However, a separate double-blind, randomized, placebo-controlled study also evaluated the long-term (1 year) effects of 1 mg anastrozole daily on BMD in older men (>60 years) with borderline or low testosterone levels and hypogonadal symptoms [45]. Although bone turnover markers were not significantly changed with anastrozole therapy, the authors reported a decrease in BMD in treated patients compared to placebo, which was noted within 6 months. They concluded that aromatase inhibition does not improve skeletal health in aging men with low or low normal T levels and may have deleterious effects on BMD. It might be advisable, therefore, to limit the use of this drug to 6 months or less.

Although anastrozole treatment increases T levels in hypogonadal men, further studies are required to assess its impact on safety and sexual function. Combining anastrozole with T can increase the effect of T replacement and may be helpful in treating hypogonadal men.

Human Chorionic Gonadotropin

LH and beta-hCG have similar molecular structures and both bind on the same receptor in Leydig cells [16]. In a recent study, which included 100 men with hypogonadotropic hypogonadism, treatment with hCG for 3–6 months restored serum T concentrations in 81 of them [46]. However, further randomized, controlled trials are needed to elucidate the role of beta-hCG in the treatment of hypogonadism. There is always the concern that the T stimulation will have a negative feedback effect on the pituitary hormones, LH and FSH, which theoretically can reduce spermatogenesis, although anecdotal evidence suggests that this effect is minimal or nonexistent. More formal studies are needed to clarify this effect. It is also to be noted that hCG is only approved for ovulation induction, and the treatment of certain hypothalamic–pituitary conditions such as Kallman's syndrome in the USA.

Conclusion

The most commonly employed treatment for male hypogonadism is exogenous T administration. However, exogenous T affects the natural mechanisms of the hypothalamo–pituitary–gonadal axis which can impair spermatogenesis and cause testicular atrophy. Although these side effects are usually reversible, they are of particular concern for younger patients and older men who are still wishing to procreate. Alternative treatment options may be considered in such cases. Further studies are required to determine their actual role in treating hypogonadal men.

References

1. Araujo AB, Esche GR, Kupelian V, et al. Prevalence of symptomatic androgen deficiency in men. J Clin Endocrinol Metab. 2007;92:4241–7.
2. Maggi M, Schulman C, Quinton R, Langham S, Uhl-Hochgraeber K. The burden of testosterone deficiency syndrome in adult men: economic and quality-of-life impact. J Sex Med. 2007;4:1056–69.

3. Davidson JM, Camargo CA, Smith ER. Effects of androgen on sexual behavior in hypogonadal men. J Clin Endocrinol Metab. 1979;48:955–8.
4. Kaler LW, Neaves WB. The androgen status of aging male rats. Endocrinology. 1981;108: 712–9.
5. Kwan M, Greenleaf WJ, Mann J, Crapo L, Davidson JM. The nature of androgen action on male sexuality: a combined laboratory-self-report study on hypogonadal men. J Clin Endocrinol Metab. 1983;57:557–62.
6. O'Carroll R, Shapiro C, Bancroft J. Androgens, behaviour and nocturnal erection in hypogonadal men: the effects of varying the replacement dose. Clin Endocrinol (Oxf). 1985;23: 527–38.
7. Jackson JA, Riggs MW, Spiekerman AM. Testosterone deficiency as a risk factor for hip fractures in men: a case–control study. Am J Med Sci. 1992;304:4–8.
8. Stoch SA, Parker RA, Chen L, et al. Bone loss in men with prostate cancer treated with gonadotropin-releasing hormone agonists. J Clin Endocrinol Metab. 2001;86:2787–91.
9. Ferrando AA, Sheffield-Moore M, Yeckel CW, et al. Testosterone administration to older men improves muscle function: molecular and physiological mechanisms. Am J Physiol Endocrinol Metab. 2002;282:E601–7.
10. Simon D, Charles MA, Nahoul K, et al. Association between plasma total testosterone and cardiovascular risk factors in healthy adult men: the telecom study. J Clin Endocrinol Metab. 1997;82:682–5.
11. Jenkins VA, Bloomfield DJ, Shilling VM, Edginton TL. Does neoadjuvant hormone therapy for early prostate cancer affect cognition? Results from a pilot study. BJU Int. 2005;96:48–53.
12. Slabbekoorn D, van Goozen SH, Megens J, Gooren LJ, Cohen-Kettenis PT. Activating effects of cross-sex hormones on cognitive functioning: a study of short-term and long-term hormone effects in transsexuals. Psychoneuroendocrinology. 1999;24:423–47.
13. Shores MM, Sloan KL, Matsumoto AM, Moceri VM, Felker B, Kivlahan DR. Increased incidence of diagnosed depressive illness in hypogonadal older men. Arch Gen Psychiatry. 2004;61:162–7.
14. Bhasin S, Cunningham GR, Hayes FJ, et al. Testosterone therapy in men with androgen deficiency syndromes: an endocrine society clinical practice guideline. J Clin Endocrinol Metab. 2010;95:2536–59.
15. Morales A, Heaton JP, Carson 3rd CC. Andropause: a misnomer for a true clinical entity. J Urol. 2000;163:705–12.
16. Zitzmann M, Nieschlag E. Hormone substitution in male hypogonadism. Mol Cell Endocrinol. 2000;161:73–88.
17. Bhasin S, Basaria S. Diagnosis and treatment of hypogonadism in men. Best Pract Res Clin Endocrinol Metab. 2011;25:251–70.
18. Calof OM, Singh AB, Lee ML, et al. Adverse events associated with testosterone replacement in middle-aged and older men: a meta-analysis of randomized, placebo-controlled trials. J Gerontol A Biol Sci Med Sci. 2005;60:1451–7.
19. Rhoden EL, Morgentaler A. Risks of testosterone-replacement therapy and recommendations for monitoring. N Engl J Med. 2004;350:482–92.
20. Guay AT, Bansal S, Heatley GJ. Effect of raising endogenous testosterone levels in impotent men with secondary hypogonadism: double blind placebo-controlled trial with clomiphene citrate. J Clin Endocrinol Metab. 1995;80:3546–52.
21. World Health Organisation (WHO). Contraceptive efficacy of testosterone-induced azoospermia and oligozoospermia in normal men. Fertil Steril. 1996;65:821–9.
22. Shabsigh A, Kang Y, Shabsign R, et al. Clomiphene citrate effects on testosterone/estrogen ratio in male hypogonadism. J Sex Med. 2005;2:716–21.
23. Rolf C, Knie U, Lemmnitz G, Nieschlag E. Interpersonal testosterone transfer after topical application of a newly developed testosterone gel preparation. Clin Endocrinol (Oxf). 2002;56:637–41.
24. de Ronde W. Hyperandrogenism after transfer of topical testosterone gel: case report and review of published and unpublished studies. Hum Reprod. 2009;24:425–8.

25. Han TS, Bouloux PM. What is the optimal therapy for young males with hypogonadotropic hypogonadism? Clin Endocrinol (Oxf). 2010;72:731–7.
26. Ko EY, Siddiqi K, Brannigan RE, Sabanegh Jr ES. Empirical medical therapy for idiopathic male infertility: a survey of the American urological association. J Urol. 2012;187:973–8.
27. Schiff JD, Ramirez ML, Bar-Chama N. Medical and surgical management male infertility. Endocrinol Metab Clin North Am. 2007;36:313–31.
28. Goldstein SR, Siddhanti S, Ciaccia AV, Plouffe Jr L. A pharmacological review of selective oestrogen receptor modulators. Hum Reprod Update. 2000;6:212–24.
29. Tenover JS, Bremner WJ. The effects of normal aging on the response of the pituitary-gonadal axis to chronic clomiphene administration in men. J Androl. 1991;12:258–63.
30. Katz DJ, Nabulsi O, Tal R, Mulhall JP. Outcomes of clomiphene citrate treatment in young hypogonadal men. BJU Int. 2012;110(4):573–8.
31. Guay AT, Jacobson J, Perez JB, Hodge MB, Velasquez E. Clomiphene increases free testosterone levels in men with both secondary hypogonadism and erectile dysfunction: who does and does not benefit? Int J Impot Res. 2003;15:156–65.
32. Ioannidou-Kadis S, Wright PJ, Neely RD, Quinton R. Complete reversal of adult-onset isolated hypogonadotropic hypogonadism with clomiphene citrate. Fertil Steril. 2006;86:1513.e5–9.
33. Taylor F, Levine L. Clomiphene citrate and testosterone gel replacement therapy for male hypogonadism: efficacy and treatment cost. J Sex Med. 2010;7:269–76.
34. Ribeiro RS, Abucham J. Clomiphene fails to revert hypogonadism in most male patients with conventionally treated nonfunctioning pituitary adenomas. Arq Bras Endocrinol Metabol. 2011;55:266–71.
35. Ribeiro RS, Abucham J. Recovery of persistent hypogonadism by clomiphene in males with prolactinomas under dopamine agonist treatment. Eur J Endocrinol. 2009;161:163–9.
36. Kaminetsky J, Hemani ML. Clomiphene citrate and enclomiphene for the treatment of hypogonadal androgen deficiency. Expert Opin Investig Drugs. 2009;18:1947–55.
37. Hill S, Arutchelvam V, Quinton R. Enclomiphene, an estrogen receptor antagonist for the treatment of testosterone deficiency in men. IDrugs. 2009;12:109–19.
38. Raman JD, Schlegel PN. Aromatase inhibitors for male infertility. J Urol. 2002;167:624–9.
39. Holbrook JM, Cohen PG. Aromatase inhibition for the treatment of idiopathic hypogonadotropic hypogonadism in men with premature ejaculation. South Med J. 2003;96:544–7.
40. Burnett-Bowie SA, Roupenian KC, Dere ME, Lee H, Leder BZ. Effects of aromatase inhibition in hypogonadal older men: a randomized, double-blind, placebo-controlled trial. *Clin Endocrinol (Oxf)*. 2009;70(1):116–23.
41. Herzog AG, Farina EL, Drislane FW, et al. A comparison of anastrozole and testosterone versus placebo and testosterone for treatment of sexual dysfunction in men with epilepsy and hypogonadism. Epilepsy Behav. 2010;17:264–71.
42. Leder BZ, Rohrer JL, Rubin SD, Gallo J, Longcope C. Effects of aromatase inhibition in elderly men with low or borderline-low serum testosterone levels. J Clin Endocrinol Metab. 2004;89:1174–80.
43. Dougherty RH, Rohrer JL, Hayden D, Rubin SD, Leder BZ. Effect of aromatase inhibition on lipids and inflammatory markers of cardiovascular disease in elderly men with low testosterone levels. Clin Endocrinol (Oxf). 2005;62:228–35.
44. Leder BZ, Finkelstein JS. Effect of aromatase inhibition on bone metabolism in elderly hypogonadal men. Osteoporos Int. 2005;16:1487–94.
45. Burnett-Bowie SA, McKay EA, Lee H, Leder BZ. Effects of aromatase inhibition on bone mineral density and bone turnover in older men with low testosterone levels. J Clin Endocrinol Metab. 2009;94:4785–92.
46. Warne DW, Decosterd G, Okada H, Yano Y, Koide N, Howles CM. A combined analysis of data to identify predictive factors for spermatogenesis in men with hypogonadotropic hypogonadism treated with recombinant human follicle-stimulating hormone and human chorionic gonadotropin. Fertil Steril. 2009;92:594–604.

Index

W.J.G. Hellstrom (ed.), *Androgen Deficiency and Testosterone Replacement: Current Controversies and Strategies*, Current Clinical Urology,
DOI 10.1007/978-1-62703-179-0, © Springer Science+Business Media, LLC 2013

MIX
Papier aus verantwortungsvollen Quellen
Paper from responsible sources
FSC® C105338

If you have any concerns about our products,
you can contact us on
ProductSafety@springernature.com

In case Publisher is established outside the EU,
the EU authorized representative is:
Springer Nature Customer Service Center GmbH
Europaplatz 3, 69115 Heidelberg, Germany

Printed by Libri Plureos GmbH
in Hamburg, Germany